Regional Productivity Growth in China's Agriculture

Regional Productivity Growth in China's Agriculture

Shenggen Fan

LONDON AND NEW YORK

First published 1990 by Westview Press

Published 2019 by Routledge
52 Vanderbilt Avenue, New York, NY 10017
2 Park Square, Milton Park, Abingdon, Oxon OX14 4RN

Routledge is an imprint of the Taylor & Francis Group, an informa business

Library of Congress Cataloging-in-Publication Data
Fan, Shenggen.
Regional productivity growth in China's agriculture / by Shenggen Fan.
p. cm.—(Westview special studies in agriculture science and policy)
Includes bibliographical references and index.
ISBN 0-8133-7989-X
1. Agricultural productivity—China—Regional disparities.
I. Title. II. Series.
HD2097.F35 1990
338.1'0951—dc20 **90-12396**
CIP

ISBN 13: 978-0-367-28549-4 (hbk)

ISBN 13: 978-0-367-30095-1 (pbk)

CONTENTS

TABLES

FIGURES

FOREWORD

This study by Shenggen Fan makes three important and original contributions. It is the first study to report regional patterns of productivity growth in Chinese agriculture. There have been dramatic differences in output and productivity growth among Chinese regions. The second contribution is to measure the separate effects of technical change and institutional reform on productivity growth. Much of the rapid growth in agricultural production and in productivity since the late 1970s has been a consequence of an important series of institutional reforms. The third contribution is the first test of the induced innovation hypothesis against experience in a centrally planned economy. Regional patterns of productivity growth are consistent with the hypothesis that the path of technical change has been responsive to regional differences in resource endowments.

Chinese agriculture has experienced a series of rapid growth "spurts" followed by periods of relative stagnation since liberation in 1949 (Table 1, p. xxiii). In spite of these cycles, output growth in Chinese agriculture has been rapid compared to other centrally planned economies and compared to all but a few of the most successful third world developing countries. Between 1980 and 1985, following the series of institutional reforms known as the "household responsibility system," which involved the transition from communal production to production managed by individual contractor households, agricultural output grew by a remarkable eight percent per year. Although there has been considerable controversy about the sources of this rapid growth, our own hypothesis has been that much of it was explained by the release of productive resources resulting from the institutional reforms. Even in advanced industrial countries, the rate of growth of output per unit of total input has rarely exceeded two percent per year. It seemed to us that the period of rapid growth following the institutional reforms of the end of the 1970s and the beginning of the 1980s would be followed by considerably slower growth once the backlog of technology whose application had been dampened by institutional constraints had been exhausted. It was not clear to us, or to other students of Chinese agricultural development, however, how long one might expect the rapid growth to continue. The evidence now available suggests a sharp decline in growth rates since the mid-1980s. Fan's study indicates that in addition to institutional reform, a very substantial part of agricultural growth in China is accounted for by the introduction of modern inputs. Rapid growth in the use of chemical fertilizers beginning in the late 1960s, when a number of large fertilizer plants came on stream, was particularly important. Other factors, such as the expansion of irrigated area, the introduction of modern crop varieties, and other industrial inputs such as pesticides and plastic sheeting (for intensive greenhouse production), have also been important.

Fan's study indicates that the potential impact of these modern inputs has not yet been exhausted. One indicator is the relatively large production elasticities for the modern inputs calculated from the statistically estimated production functions. Furthermore, there is a substantial gap between factor shares to land and labor calculated from accounting data and production elasticities measured from the production functions. This suggests that increased use of the modern inputs will contribute not only to growth of agricultural output, but will reduce costs of production through continued reallocation of resources between modern and traditional inputs. Growth in productivity as well as total output is necessary if the agricultural sector is to provide food and fiber to a growing population at politically acceptable costs. The ease with which this substitution can take place in agriculture will depend on growth in non-agricultural employment for rural people. Fan's interregional comparisons show that the increase in total factor productivity was greatest where opportunities for excess labor to move out of agricultural production were also great.

Since the mid-1980s institutional constraints appear to have slowed this substitution of modern for traditional inputs. Expansion of irrigation capacity and of irrigated area has slowed. The reorganization of local governance associated with the reforms of the late 1970s has disrupted the capacity of townships and higher levels of local government to maintain physical and institutional infrastructure. The level of public investment allocated to agriculture is shockingly low and poorly distributed regionally. The marketing infrastructure required for a market economy remains underdeveloped. As production has slowed, the authorities have re-instituted the requisition of grains at well below market clearing price levels. Incentives to producers have been dampened both on the factor market and product market side. The challenges that face the Chinese agricultural economy during the next several decades are immense. Population is expected to grow from approximately one billion in 1980 to somewhere near 1.25 billion by the year 2000. This implies that somewhere in the neighborhood of 550 million metric tons of grain will be needed by the turn of the century--an increase of 150 million metric tons over the level achieved in the mid-1980s.[1]

Achievement of increases in production of this magnitude will require rapid growth in productivity generated both by technical change and continuing institutional reform. These reforms will include the development of modern market institutions capable of delivering inputs and transmitting products to consumers efficiently and of providing strong incentives to producers. It is highly unlikely that the objectives can be achieved if the government continues to involve itself as directly as at present in determining market prices and in the logistical aspects of input supply and product marketing.

It will also require substantial strengthening of the Chinese agricultural research system. The Chinese agricultural research system has been remarkably productive in spite of limited scientific capacity. In some sense, it has substituted large numbers of inadequately trained professional and scientific staff for the higher quality human capital in agricultural research. This will have to change in the future. This means continued training of Chinese agricultural scientists, both at home and abroad, and provision of adequate incentives to attract them to work in China on important agricultural problems.

It will also be necessary, over the next several decades, to narrow the disparities in productivity and output growth among regions. If this is to occur, it will be necessary to make the investments in transportation and other marketing facilities that will strengthen inter-regional movement of inputs and products. Additional regional specialization should become a more important source of productivity growth than it has been in the past.

While this study has provided an initial framework for thinking about regional productivity growth in China, it does no more than point the way to the future research that should be done. We list the following among the questions that are important from both a policy and a scholarly perspective.

1. Why was institutional reform initially successful in China, and not in other socialist countries? What can the other socialist countries learn from China? Is the Chinese model transferable?

2. More micro-level analysis should be conducted. What are the sources of efficiency or productivity differences among households? How does the response to the government policies, new technology, and education vary among households?

3. What are the effects of macro-policies and sector policies in other industries on agricultural productivity growth?

4. What factors account for and will contribute to growth of non-agricultural employment opportunities for rural people?

Karen M. Brooks
Assistant Professor
University of Minnesota

Vernon W. Ruttan
Regents Professor
University of Minnesota

Table 1

STAGES IN THE TRANSFORMATION OF THE CHINESE ECONOMY

		Annual Rates of Growth in:		
		Agricultural output	Industrial output	National Income
1949-52	Economic rehabilitation; land to the tiller land reform.	14.1	34.8	19.3
1952-57	First Five-Year plan; agricultural mutual aid teams and elementary producer cooperative organized, income distribution based on work points.	4.5	18.0	8.9
1958-62	Great Leap Forward; agricultural producer cooperatives reformed into communes, production brigades and production teams; work points replaced by equality in distribution	-6.0	-6.0	-8.5
1962-65	Readjustment and Recovery. Production team as basic unit for production and distribution, private plots and free markets restored.	11.1	17.9	14.7
1965-75	Cultural Revolution	4.3	9.2	6.9
1976-80	Reform and Consolidation. Fifth five year plan was 76-80; "Gang of Four" overthrown in October 1976; liberalization began to move forward rapidly after Third Plenum of the Eleventh Central Committee of PRC in December 1978.	3.3	9.2	6.0
1980-85	Resumption of Economic Growth. Contract-responsibility system; family re-established as basic unit of production.	8.0	10.8	9.8
1985-88		4.6	14.7	8.6

Sources: China's Statistical Yearbook, 1987 (Beijing: The Statistical Press)
China's Statistical Abstract, 1988 (Beijing: The Statistical Press)

Note: Agricultural Output excludes the output of rural industry.

Endnote

1. Niu, Ruofeng, "China's Grain Production Toward 2000," in John W. Longworth (ed.), China's Rural Development Miracle: With International Comparisons (Queen's St. Lucia: University of Queensland Press, in association with the International Association of Agricultural Economists and the Australian Development Assistance Bureau, 1989), pp.189-195.

MINNESOTA STUDIES ON AGRICULTURE IN CENTRALLY PLANNED ECONOMIES

The research reported in this book is part of a continuing series of studies on agriculture in the centrally planned economies being carried out at the Center for International Food and Agricultural Policy and the Department of Agricultural and Applied Economics at the University of Minnesota.

Books

Jean Sussman, Agricultural Research in Cuba (Boulder: Westview, forthcoming).

Long Fai Wong, Agricultural Productivity in the Socialist Countries (Boulder: Westview, 1986).

Recent Papers and Chapters

Karen M. Brooks, "The Law on Cooperatives, Retail Food Prices, and the Farm Financial Crisis in the USSR," Staff Paper p88-29, Department of Agricultural and Applied Economics, University of Minnesota.

Karen M. Brooks, "Agricultural Reform in the Soviet Union," in Carl K. Eicher and John M. Staatz, eds., Agricultural Development in the Third World (Baltimore: Johns Hopkins University Press, 1989, second edition).

Lung Fai Wong and Vernon W. Ruttan, "Sources of Differences in Agricultural Productivity Growth Among Socialist Countries" in Ali Dogramaci and Rolf Fare (eds.), Applications of Modern Production Theory: Efficiency and Productivity (Boston: Kluwer Academic Publishers, 1988), pp. 103-130.

Lung Fai Wong, "Agricultural Productivity Growth in China and India: A Comparative Analysis," Canadian Journal of Agricultural Economics 37 (1989), pp. 77-93.

Thesis

Masahiko Gemma, Productivity Growth in Polish Agriculture: A Comparative Study (Ph.D. Thesis, Graduate School of University of Minnesota, 1989).

ACKNOWLEDGMENTS

I wish to express my sincere appreciation to Dr. Vernon W. Ruttan, Regents Professor at the University of Minnesota. He spent a great deal of invaluable time to read the manuscript at various stages and returned it to me with helpful suggestions. This book could not have been completed without his guidance and encouragement.

I have benefited from comments and suggestions from many faculty at the University of Minnesota. They include Dr. G. Edward Schuh of the Hubert H. Humphrey Institute; Dr. Willis L. Peterson and Dr. Karen M. Brooks of the Department of Agricultural and Applied Economics; and Dr. J. Werner of the Economics Department.

Professor Ruofeng Niu at the Institute of Agricultural Economics of the Chinese Academy of Agricultural Sciences, and Professor Wenpu Liu, Mr. Yalai Zhang, and Mr. Xiaoshan Du at the Institute of Agricultural Economics of the Chinese Academy of Social Sciences, arranged the data collection in China for me. To them, I am indeed thankful. My appreciation is also extended to Sylvia Rosen for her editorial assistance.

Financial support for this research was provided by grants to the University of Minnesota from the Rockefeller Foundation and Minnesota Experimental Station, for which I am grateful.

CHAPTER 1
INTRODUCTION

A shift from traditional to modern patterns of farming is currently transforming agriculture in less developed countries. The changes are both institutional and technological in content, that is, agricultural production is increasingly organized in an integrated system and uses more modern inputs (or technical inputs) on a large scale provided by off-farm sectors. This transformation has brought about rapid production growth and increased food availability for the world population. One of the countries in which this shift has taken a dramatic turn is the People's Republic of China. The growth of agricultural production there has been high; indeed, between 1949 and 1986, the growth rate was 4% per year. This achievement is accepted as fact, but there are no answers to questions of what productivities and efficiencies made it possible, what are the regional differences of this performance, and how to promote further production growth. This book attempts to fill this gap in our knowledge.

Traditional agriculture has been identified always by the organization of production for self-sufficiency in which available resources are allocated efficiently and the rate of return to investment at the margins is low (Schultz, 1964). The patterns may be seen in any of a variety of institutional forms, from highly communalized systems to small farms organized around family units (Mellor, 1966). In China, as in other developing socialist countries where agriculture was organized into collectives or state farms, the patterns of production were traditional although large in scale, and farming activities and family consumption were intermixed.

Socialist and nonsocialist agricultural systems differ sharply in terms of who makes the production decisions and what are goals of production. In nonsocialist agriculture, farmers make their own decisions on what to grow and what resources to use; the basis of such decisions is to try to equalize marginal revenues among different activities. In socialist agriculture, on the other hand, production decisions are made by team leaders or managers; they allocate resources and set production goals on the bases of first fulfilling the government's national goals by meeting production targets and then maximizing profits. Consequently, cultivation activities are not necessarily the most efficient. Nevertheless, if we consider such production targets to be given or inviolable constraints, then the operations of the collectives or state farms can be interpreted as conforming to profit maximization theory insofar as team leaders or managers allocate resources to equalize marginal returns among the different production activities.

Traditional agriculture, whether practiced on large- or small-scale production units, also is characterized by low productivity and slow growth. Both economic theory and empirical studies support the notion that traditional agriculture is caught in a technical and economic equilibrium trap (Stevens, 1977). To escape from this trap, a greater effort is needed to change producers' use of traditional farming techniques and to encourage the use of more modern production inputs, for example, chemical fertilizers, machinery, and technical education, which can be produced or provided by other sectors of the economy.

An initial low productivity growth stage in the modernization of agriculture seems to be inevitable. The United States went through it from 1880 to 1920 and Japan, from 1920 to 1960 (Hayami and Ruttan, 1971/1985), and most developing countries are going through the process currently. When Wong (1986) studied nine socialist countries and India, he found agriculture to be struggling through the process in those nations.

In China, modernization efforts may be having a negative effect on total agricultural productivity. Government units assign individuals to jobs and, perhaps more important, control the internal migration of people from countrysides to cities and towns. Thus, increased access to more production inputs creates a potential conflict between agricultural productivity and labor. The greater the amounts of modern inputs, the fewer the people who are needed to increase production. But this labor surplus has no place to go. The industrial sector is not able to absorb them and they do not add to agricultural productivity. Indeed, when the land/labor ratio becomes unfavorable, low or negative total factor productivity growth occurs.

What happens in Chinese agriculture is of international interest because of its contribution to the global economy. Although the nation has only about 7% of the world's arable land, it feeds about 23% of the world's population. So what is going on there? During the decade of the 1980s, agricultural development in China displayed three distinct characteristics: (1) its agriculture moved from traditional to modern patterns; (2) it successfully modeled the possibilities of agricultural reform in a socialist country and, hence, was able to make tested contributions to both developing and socialist countries; and (3) its land-saving technology prevailed throughout agriculture just as it does in most Asian countries.

The sources of growth in China's agricultural production are not well understood, however. We know that increasing agricultural production depends upon rising factor inputs and productivity growth, but in China, the essential factor inputs are limited. Arable land areas have been decreased by expanding cities and spreading industrialization whereas agricultural labor has been increasing steadily because of rapid population growth and restricted migration. Yet, while production was increasing, the ratio of labor to land was becoming very unfavorable.

When scarcity of land is the main constraint on agricultural growth, farming must become labor intensive and require more and more inputs, such as chemical fertilizer, high-yield varieties, and other sophisticated materials. At the same time, land productivity can be increased by improvements in regional infrastructure, such as large-scale water conservation and capital investments on farms. The rapid increase of machinery input, it has been found, contributes little to labor productivity. Given the fixed quantity of arable land in China and its labor-surplus economy, any increase in labor productivity necessarily derives from land productivity growth. The extent to which such changes affect efficiency can be ascertained only by quantitative analyses.

What sets China apart from most other developing countries is not that China has increased the use of technical inputs, but, rather, how much it has relied on organizational reform to increase agricultural production (Perkins and Yusuf, 1984). When the Communists took power in the late 1940s, one of their primary objectives was the confiscation of land from landlords and redistributing it to poor peasants (Pannell and Ma, 1983). During the second half of the 1950s, the large-scale construction of People's Communes was carried out. It was these communes and their subunits--the production brigades and production teams--that became the essential units of agricultural production and organization until the recent institutional reforms. Beginning with 1979, efforts were made to create agricultural incentives and to promote production by decentralizing authority and increasing the responsibility of family units. The large-scale introduction of the family household responsibility system not only contributed to recent rapid growth in agricultural production but also resulted in the rethinking of economic theory in socialist agriculture. For these reasons, comparing the efficiencies in different institutional systems permits us to make more secure judgments on how institutions affect production and productivity growth.

China's diverse climate and topography result in huge differences in culture, economic development, and resource endowments and, hence, in diverse farming techniques, cropping patterns, and productivities. Of the country's more than one billion people, 90% are concentrated east of a line running from Aihui County in Heilongjiang Province southwest to Tengchong County in Yunnan Province. This area incorporates about 90% of the country's arable land, forested areas, and inland waters. People are scarce in West China; most are minorities who live in the dry regions in compact ethnic communities. It can be truthfully stated, therefore, that of China's 9.6 million square kilometers of territory, only the eastern half actually functions to feed the more than one billion inhabitants. Yet, even in the eastern part of the country, the differences in cropping patterns are very large. The use of national aggregate data, consequently, conceals different kinds and sources of productivity change.

Chinese scholars have tended to pay little attention to the measurement of productivity. The national aggregate data analyses reported by some scholars were inconclusive. For example, Tang (1984) found total factor productivity to decrease from .3% to .5% during the period 1952-1980 whereas Wong (1986) came up with greater decreases. After constructing a total factor productivity index for the nine socialist countries he studied, Wong concluded that in China, total factor productivity declined significantly from 1950 to 1980 (5.65% a year using the arithmetic index approach and 3.14% a year using the geometric index approach).

Other investigators came up with different conclusions. Rawski (1982) estimated total factor productivity in China to have deteriorated at the rate of 1.7% to 2.4% a year from 1957 to 1975, but Wiens (1982), who studied technical change in China's agriculture, concluded that total factor productivity grew at .8% a year from 1957 to 1978. In Yamada's (1975) comparison of Asian agricultural productivity, he attributed to China's agriculture a 1.58% annual growth of total factor productivity from 1963 to 1970.

Perkins and Yusuf (1984) claimed that the signs and magnitude of total factor productivity growth depend on the estimation of the marginal product of rural labor. *"If the marginal product of rural labor was high, productivity growth per unit of input were most likely negative. Only if it is argued that the mass mobilization of rural labor for collective activities had only modest effect on output, can a plausible case be made that factor productivity in Chinese farming was rising."* (Page 68). Their measurement of total factor productivity in Chinese agriculture ranged from -.13% to .27% a year from 1965 to 1979.

The inconsistent results arise mainly from differences in the measurement of inputs and outputs, and from the weights adopted to aggregate the indexes for total input and total output. In addition, technical change is considered to be a unique source of productivity change for both agriculture and non-agriculture sectors in the literature. This assumption, however, does not take account of the effect of efficiency on the improvement of productivity growth, given the same technology. Indeed, the definitions of technical change and efficiency improvement tend to be misleading because far too often the two are used synonymously.

The comparison of productivity growth among the regions of a country has not been given enough attention in the literature of production study which means that the sources of regional productivity differences are not well understood.

A number of questions on Chinese agriculture were answered in this book; first and foremost was the question of how China managed to achieve the high growth rate of agricultural production. To answer this question it was necessary first to determine the following:

1. Given the size of China and the differences among regions, how do the differences in natural resource endowment and in social and economic conditions among regions affect agricultural development?

2. What are the physical relations between inputs and outputs in Chinese agriculture? How do they change over time?

3. What is the productivity growth pattern for each region? Have the productivity differences among regions increased over time?

4. What are the main sources of productivity growth over time and the main sources of productivity differences, if any, among regions?

5. Does the induced innovation theory work in a non-market economy such as China's?

The book begins in Chapter 2 with an overview of the regional characteristics of natural resources, social and economic conditions, and agricultural development, which provide the essential background for the succeeding chapters. The chapter can also be read as an independent comparison of regional features.

Frontier production functions are estimated in

Chapter 3. The functions are first specified, the choice of functional forms is discussed, and then the coefficients of the functions and average efficiency are estimated. Finally, the results are compared with those of previous studies.

Based on the measurement of partial productivities, and the theoretical discussion and measurement of total factor productivity, the trends of productivities are examined in Chapter 4. A comparison of productivity growth patterns among regions concludes the chapter.

In Chapter 5, an account is given of the sources of productivity change over time. More important, the sources of regional productivity differences are analyzed. Analyses for both 1965 and 1986 are conducted in order to observe the change in the sources of productivity differences over time.

In the last chapter, all the findings in this study are summarized and some policy implications are presented.

CHAPTER 2
REGIONAL CHARACTERISTICS OF CHINESE AGRICULTURE

2.1 Introduction: Classification of Chinese Agricultural Regions

Since the 1950s, Chinese scholars, following the political division, have divided the country into 6 regions (see Map 1): (1) Northeast (N.E.) region of Heilongjiang, Liaoning, and Jilin provinces; (2) North (N.) region with the cities of Beijing and Tianjin; Hebei and Shanxi provinces; and Nei Monggol autonomous region; (3) Northwest (N.W.) region of Shaanxi, Gansu, and Qinghai provinces, and Ningxia Hui and Xinjiang Uygur autonomous regions; (4) Central and South (C.S.) region of Henan, Hubei, Hunan, Guangdong provinces, and Guangxi autonomous region; (5) East (E.) region of Shanghai city, Jiangsu Zhejiang, Anhui, Fujian, Jiangxi, and Shandong provinces; and (6) Southwest (S.W.) region of Sichuan, Guizhou, and Yunnan provinces, and Tibet autonomous region. A recent mapping of regions by agricultural and natural conditions resulted in a more precise and more detailed division of the country, but it is not consistent with the data.

Map 1. Chinese Agricultural Regions (Old)

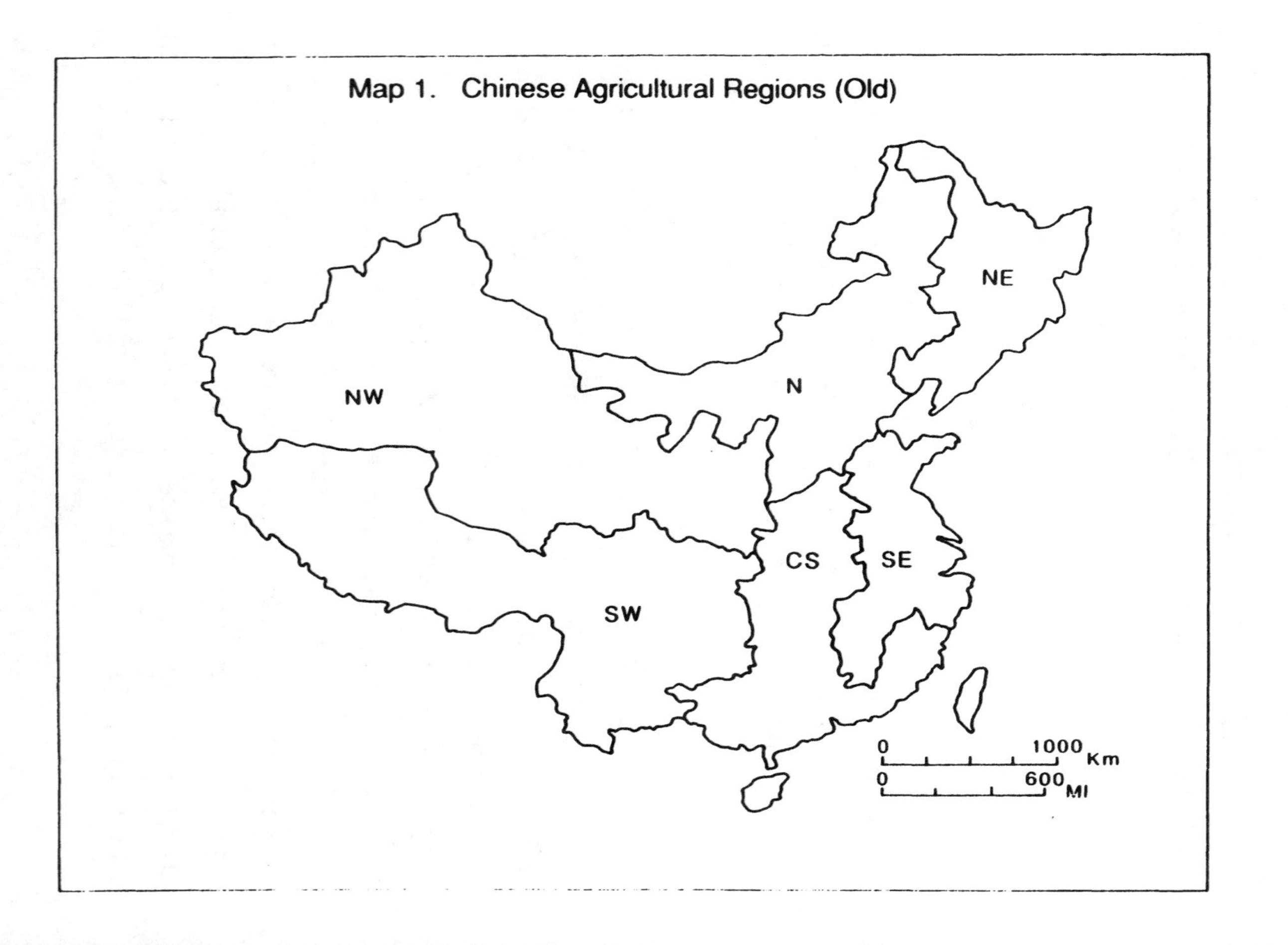

Taking into account the consistency of the data, the agricultural geographical features, and the current social and cultural conditions, we divide China into 7 regions for analytic purposes. Thus, we place Tibet and Nei Monggol in the Northwest region because livestock husbandry predominates in both regions. Gansu, Shaanxi, Shandong, and Henan are classified as the North region given that they have the same type of agriculture: a predominance of crop production in which irrigation has an important role (we could call it "irrigation agriculture"). The Central Southern region is divided into two regions: South, which has tropical agricultural features, and Central, in which rice and water products are important.

The seven regions (see Map 2), then, are (1) Northeast (N.E.), including Heilongjiang, Liaoning, and Jilin Provinces; (2) North (N.), made up of the cities of Beijing and Tianjin, and Hebei, Henan, Shandong, Shanxi, Shaanxi, and Gansu provinces; (3) Northwest or Border Region (N.W.), comprising the autonomous regions of Nei Monggol, Ningxia, Xinjiang, and Tibet; and Qinghai province; (4) Central (C.), made up of Jiangxi, Hunan, and Hubei provinces; (5) Southeast (S.E.), which includes heavily populated Shanghai city and Jiangsu, Zhejiang, and Anhui provinces; (6) Southwest (S.W), comprising Sichuan, Guizhou, and Yunnan provinces, and (7) South (S.), embracing the coastal provinces of Fujian and Guangdong province, and Guangxi autonomous region.

Map 2. Chinese Agricultural Regions (New)

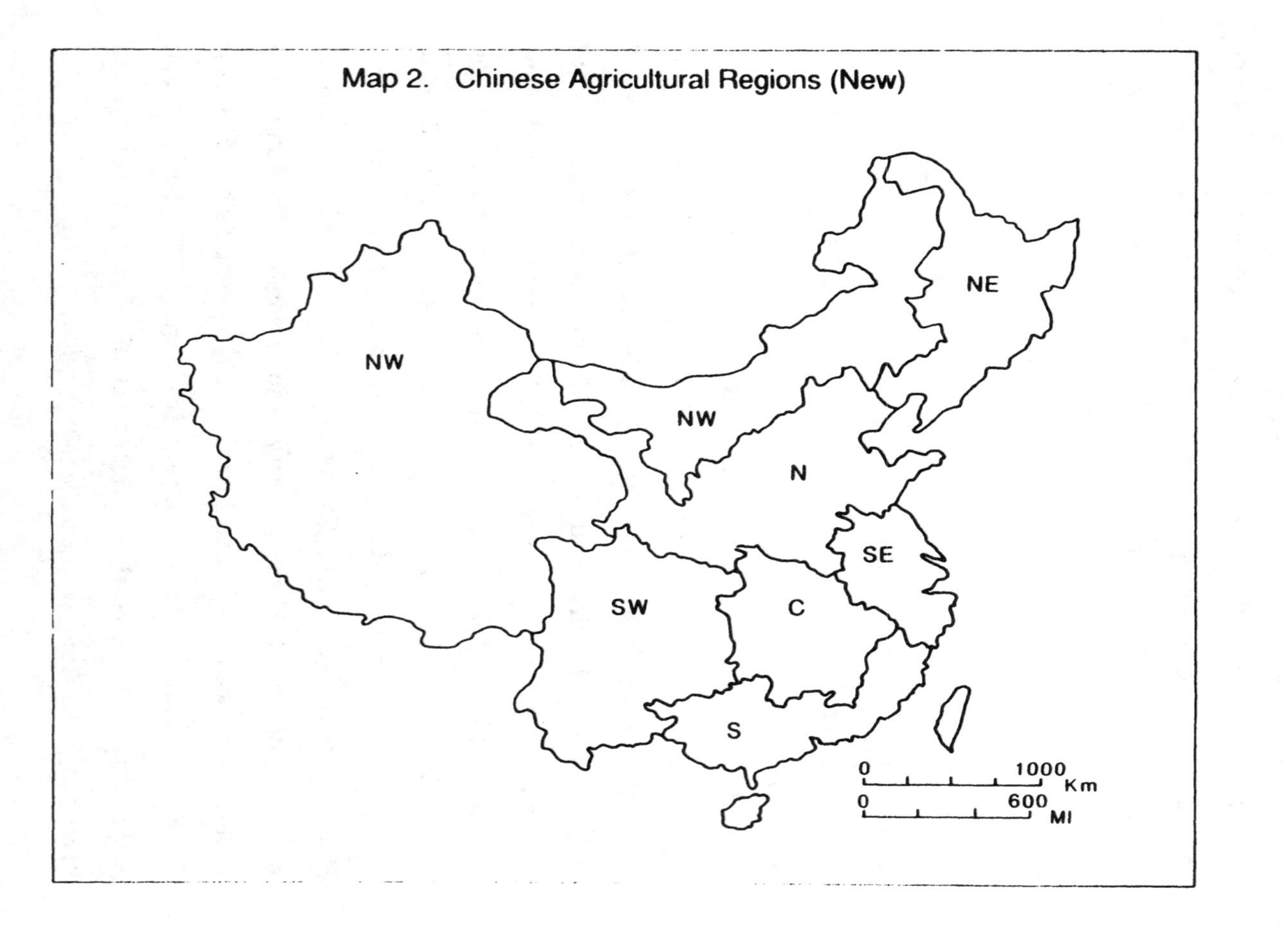

2.2 Regional Characteristics of Natural Resources

China's agricultural resources are defined by mountainous topography, unequal distribution of water resources, and various types of soils that determine the differences in farming systems across the seven regions.

Mountainous Topography

In terms of territory, China is the third largest country in the world with a land mass of 9.6 million square kilometers. Nevertheless, its agricultural and especially arable land is limited (Table 2.1). About 66% of the land area consists of mountains and only 11% of the total area is arable (Chen and Hu, 1983). Most of China's western area is mountainous with high plateaus. In the South and Southwest, hills predominate. It is only in the east that we find low-elevation plains.

The high altitudes in the mountainous areas and resulting low temperature keep the growing season short. Furthermore, the extremely cold weather puts constraints on crop planting, forestry, and animal husbandry, and the precipitous slopes and thin soil layers, prevent the expansion of land areas and farming. Given the rough topography, soil erosion is a constant threat to the ecology and the construction of irrigation systems and the use of mechanization are virtually impossible. Transportation in the mountainous areas is poor, which adds one more hindrance to the development of agriculture there.

Unequal Distribution of Water Resources

Rainfall is abundant but distributed unequally among regions (see Table 2.1). Both the south and east of China lie in the path of monsoonal winds so that some areas are classed as wet and others, semi-wet. The wet region comprises 32.25% of the country's territory; its aridity index (the ratio of possible evaporation to precipitation when average daily temperatures are greater than 10^{0}C) is less than 1.0. Irrigation is used only to flood rice paddies and not for dry-land crops. The semi-wet region covers 17.8% of the country's territory and its aridity index ranges from 1.0 to 1.49. Dry-land crops grow here, too, without irrigation although seasonal deficiencies of water often occur. The wet and semi-wet regions comprise just about half the country's total land.

The semi-dry regions (parts of North, Northwest, and Northeast regions) comprise 19.2% of the total land area and have an aridity index of 1.5 to 1.9; crop yields are low and unstable without irrigation. Dry regions, (mainly the Northwest) extend over 30.8% of the total land area and are semi-desert; the aridity index there ranges from 2.0 to 3.99 and agriculture, consequently, is extremely unstable. In the desert areas, the aridity index exceeds 4.0 and agriculture is possible only with irrigation.

The dividing line between the wet and dry regions extends from Daxinannin in Heilongjiang province to Changdu in Yunnan province (through Zhangjacou, Yulin, and Nanzhou). All of the South, Southeast, Central, and most of the Southwest, North, and Northeast regions lie to the

Table 2.1 Annual Characteristics of Natural Resources in Different Regions

	Annual Average Temperature (C^0)	Precipitation (millimeters)	Agric. Land (0,000 Ha)	Arable Land (0,000 Ha)	Predominant Soil Types
N.E.	4.8	821.9	6338.4	1245.4	Chernozem, and black earth
N.	11.5	721.0	5922.3	3329.0	Alluvial soil, and loess dust blown
N.W.	6.0	242.9	23067.0	1026.0	Steppes, desert, and semi-desert lands
C.	16.9	1029.9	2675.0	972.0	Red and yellow podzolic soils
S.E.	15.9	1672.6	1736.0	1130.0	Yellow podzolic soils
S.W.	15.7	911.8	4380.0	1131.0	Purple and brown soils
S.	21.6	1706.0	2605.0	715.0	Acid red and yello lateritic soils
National			50705.0	9950.0	

1. Sources: Temperature and Precipitation: China's Statistical Yearbook, 1986 (Beijing: Statistical Bureau, 1986). Agricultural land and arable land: Handbook of Economic for Technology (Shenyang: Liaoning People's Press, 1983).

2. "Agricultural Land" includes arable as well as grass lands. Statistics for grass land do not include grass lands in all regions, only those in main grassland regions. The figure given here underestimates the agricultural land.

3. Temperature and precipitation statistics are those recorded for following cities: N.E., Changchun; N., Beijing; N.W., Urumqi; C., Wuhan; S.E., Shanghai; S.W.; Chengdu; and S., Guangzhou.

east of the line. The differences in moisture between east and west determine the nature of Chinese agriculture. East of the line, water resources are abundant (annual precipitation, 400 to 2000 millimeters) and the agricultural regions comprise 92.3% of the nation's arable land and 90% of forests. In the Northwest dry and semi-dry regions, water resources are scarce (annual precipitation less than 400 millimeters) and animal husbandry is the major agricultural activity.

Various Types of Soils

The soil types also show great differences in China. Generally, they can be differentiated by climate and vegetation, that is, between the high or dry western and northwestern regions and the humid central and southeast (Pannell and Ma, 1983). In the dryer northern and western parts of China, the two types of soil are: the very thin and poor mountain soils found on the Tibetan Plateau and the prairie, steppe, and desert soils in the regions of Xinjiang, Nei Monggol, and the Loess Plateau, and parts of Northeast China. Eastern China is humid; its soils are generally podzols and are distinguished by the amount of moisture and the temperature regimes in which they develop. They form the most important agricultural soils in China. Although not very fertile, probably, they can be worked easily. To increase their fertility Chinese farmers add to them various organic manures, grasses, pond gucks, and other biodegradable materials (this soil classification is taken from Pannell and Ma, 1983).

Unfavorable Combination of Natural Resources

Crop growth not only needs abundant resources but also, and more important, appropriate combinations. In China, in all regions, the combinations are unfavorable to agricultural production. The Northeast has large plain areas, highly fertile soils, abundant water and forestry resources; the low temperatures and short growing season, however, constrain the further increase of agricultural production. The largest plain is in north China, a region that is hot in summer and cold in winter, with limited and unevenly distributed water resources, which make the region unfavorable to crop growth. Also, large areas of North China suffer from drought, floods, or saline-alkalization. The vast territory of the Northwest has sufficient sunlight and high temperatures in summer but agriculture is constrained by limited water resources, desert, large saline-alkali areas and extreme cold in winter. In the South, heat and water are abundant, crops grow quickly, and biological resources are plentiful, but the proportion of hill and mountain areas is much greater than that of the arable areas. Furthermore, the variation in precipitation and frequent flooding block further increases in agricultural production.

Different Farming Systems

These different natural conditions determine the farming systems instituted from North to South and from East to West. Starting in the north and moving southwards, cultivation is distinguished by four systems: (1) single crop, spring-grown temperate cereals in the northeast; (2) a winter wheat/summer crop cycle (three crops in two years) in the North China Plain; (3) double cropping with a summer rice crop in the Changjiang basin of Central China; and (4) double (occasionally triple) cropping in the tropical southern coastal area (World Bank, 1982).

2.3 Social and Economic Conditions for Agriculture

An increasingly important role in agricultural development, especially during the transition from traditional to modern agriculture, is played by a country's social and economic conditions. Traditional agriculture depends on the exploitation of resources or the interior cycle of resources. Modern agriculture attains high yields by using modern inputs that are produced away from farms. Thus, urban centers and industry are critical to agricultural production. But population, industry, and urban areas are not well distributed geographically in China (see Table 2.2); the effects on agricultural growth and development, therefore, are significant.

Unequal Distribution of Population

In the different regions of China agriculture reacts to the size of the populations in terms of both demand and supply. A large population requires a large supply of food and other agricultural products. In traditional agriculture with its poor rural infrastructure,

Table 2.2 Social and Economic Indicators for Different Regions, 1985

	Population (0,000)	Agr. Population (0,000)	Total Labor (0,000)	Rural Labor (0,000)	GPV (00 million Yuan)	APV (00 million Yuan)
N.E.	9,295	3,462	3,972.5	1,758.7	1,737.9	397.3
	(8.9)	(5.2)	(8.0)	(4.7)	(13.0)	(8.7)
N.	30,394	18,491	14,352.3	10,928.5	3,722.0	1,294.4
	(29.1)	(27.9)	(28.8)	(29.5)	(27.9)	(28.3)
N.W.	4,389	2,628	1,866.5	1,109.6	411.5	173.6
	(4.2)	(4.0)	(3.7)	(3.0)	(3.1)	(3.8)
C.	14,013	9,441	6,598.3	5,113.1	1,468.3	583.1
	(13.4)	(14.2)	(13.2)	(13.8)	(11.0)	(12.7)
S.E.	16,616	11,762	8,896.2	6,705.6	3,415.0	1,037.9
	(15.9)	(17.8)	(17.8)	(18.1)	(25.6)	(22.7)
S.W.	16,562	12,187	8,145.6	6,788.6	1,236.9	522.6
	(15.8)	(18.4)	(16.3)	(18.3)	(9.3)	(11.4)
S.	12,839	8,316	6,041.0	5,007.8	1,366.0	571.5
	(12.3)	(12.5)	(12.1)	(13.5)	(10.2)	(12.5)
Nation	104,532	66,288	49,872.7	37,065.1	13,336.0	4,580.3
	(100)	(100)	(100)	(100)	(100)	(100)

1. Sources: China's State Statistical Yearbook, 1985. (Beijing: State Statistical Bureau).

2. Rural population is defined as all population in rural areas. Rural labor is defined as the labor engaged in agriculture, rural industry, state enterprise and public service.

3. Agr. Population denotes agricultural population; GPV denotes gross production value (includes production value of agriculture, industry, transportation and so on); and APV denotes agricultural production value, which is the aggregation of value of crop production, forestry, animal husbandry, fishery, rural industry and sideline.

4. The numbers in parentheses are percentage of the national total.

agricultural products are provided mainly by nearby regions and interregional trade is very limited. At the same time, labor makes up a large share of the production inputs, so a large population means a large supply of labor.

Today, the vast majority of the Chinese people are concentrated in those few areas that offer the necessary environmental conditions for intensive agriculture. These areas are the North China Plain (part of the North region); the middle and lower Chang Jiang River basin and the river delta (Southeast and Central regions); the Sichuan Basin (Part of the Southwest region); and the Zhu River delta (part of the South region). These areas have relatively abundant precipitation, level topography, and the fertile alluvial soils produced by river systems. Furthermore, in the south, the Changjiang (Yangtze) and its several tributaries are navigable, which facilitates trade. Most of China's major cities (Shanghai, Beijing, Tianjin, Nanjing, Gouzhou, Changdu, and Zongqin) are found in these four fertile regions. Secondary population concentrations with lower density are found throughout the rugged uplands of much of south China, in the Songliao River Basin in the Northeast, where the climate is colder and the marshlands are extensive, along the Hexi Corridor in Gansu Province, and in the oases in Xinjiang Uygur Autonomous Region, higher population concentrations have been discouraged by the environment.

In Nei Monggol, Qinghai, Xinjing of Western China, where minority nationalities reside, the insufficient rainfall, extreme temperatures, and poor soils limit the development of agriculture as well as population size. In such areas, pastoral nomadism predominates (Pannell and Ma, 1983).

Imbalance in the Development of Industry

The imbalance in the development of industry, which is characteristic of China, also affects regional agricultural distribution. The largest leading industrial regions and cities are established in the coastal regions. In addition to high population density, high levels of urbanization, and the best transportation networks communication systems in the country, these regions provide extensive fertile farmland. Three areas always have been regarded as key industrial regions: the Northeast, North China, and lower Changjiang region. In contrast, the mountainous and arid areas in the Northwest and Southwest inhabited by minorities are industry poor (Pannell and Ma, 1983).

The imbalance in industrial development affects agricultural growth in four ways: (1) well-developed industry supplies the high-quality modern inputs that are necessary to increase agricultural production in a region; (2) the greater urbanization in industrialized regions offer a vast market for agricultural products; (3) high technology and well-developed infrastructure in industrial regions encourage the use of sophisticated production techniques and induce rapid technical changes; and (4) the industrial sectors of industrial regions absorb large numbers of surplus labor from agriculture, leading to improvement in land/labor ratio, hence raising labor productivity and total factor productivity.

Low Education Levels

Although China has a long literary tradition and shows a great improvement in literacy since 1949, the country suffers from a shortage of skills at all levels of education; in fact, the illiteracy rate currently is about 21% (from the Population Survey, 1987, Beijing: State Statistical Bureau). Even in those regions that are relatively developed, the low level of education is a very important constraint on economic development. For instance, in Jiangsu, one of the most developed regions in both industry and agriculture, the illiteracy rate (22.7%) is higher than the national average. Higher education in China is even worse than in other developing countries. College graduates comprise only .7% of the population over 25 years old (the comparable proportions are 1.1% for India, .9% for Bangladesh; World Bank, 1982). A low education level leads to a poor quality of rural labor which limits the adoption of new technology and efficient allocation of resources.

2.4 Regional Characteristics of Agricultural Development

The characteristics of natural resources and social and economic conditions in the different regions have determined agricultural development patterns. The measurement of input and total agricultural production growth in different regions is discussed and the differences in agricultural development are examined in this section.

Differences in Modern Input Applications

Agricultural technology, for example, fertilization, irrigation, mechanization and the introduction of high-yield varieties, and other biological techniques, have made great contributions to Chinese agricultural development. Unfortunately, large differences are found in their use in the various regions. Fertilizer application among the regions are shown in Table 2.3. The distribution of chemical fertilizer is used by the government as an instrument of state policies on crop production and procurement with cotton usually accorded the highest priority. In addition, the government tries to maximize the benefits of fertilizer application by favoring certain areas; generally, the areas are those with high yields in priority crops and better irrigation facilities. This preferential policy has contributed to the widespread adoption of fertilizer-responsive, high-yield varieties of rice in the South since the 1960s (Hsu, 1982).

Increased use of machinery depends on both supply and demand factors, but only the demand factors are discussed here. The primary objective for agriculture in China, a land scarce and labor abundant country, is to raise total output sufficiently to satisfy the increasing demand for

Table 2.3 Differences in Input Application among Regions

	Year	N.E.	N.	N.W.	C.	S.E.	S.W.	S.	National
C. Fertilizer (Kg per mu)	1965	.43	.60	.23	1.05	1.40	.77	2.93	.99
	1986	7.35	8.50	3.06	8.77	11.62	8.23	12.07	8.71
M. Fertilizer (Kg per mu)	1965	5.90	6.41	16.18	7.50	7.32	14.01	10.50	8.73
	1986	7.02	9.84	26.49	10.11	8.64	19.86	14.74	12.47
Power (Hp per labor)	1965	.51	.25	.83	.22	.12	.20	.23	.24
	1986	2.67	1.66	2.29	1.00	1.20	.57	.95	1.25
Irrigation (Rate %)	1965	5.40	15.50	33.20	34.10	28.70	20.40	33.10	22.50
	1986	13.20	33.50	35.50	33.60	34.30	22.10	32.00	29.90

1. Sources: China's Statistical Yearbooks, various issues, Beijing: State Statistical Bureau. Agricultural Yearbooks, various issues, Beijing: Agricultural Publishing House. China's Rural Statistical Yearbooks, various issues, Beijing: State Statistical Bureau.

2. Fertilizer is measured in pure nutrients; the land areas are the sown areas. For the measurement of manurial fertilizer, see Appendix 1.

3. Power is measured as the aggregation of machinery horsepower and the horse units of draft animals.

4. The total land used to calculate the ratio of irrigated land is the arable land. Arable land data after 1979 is not available but since arable land is a very stable resource, the figure used for 1986 is that compiled for 1979.

food engendered by population growth. Total output increase can be achieved only by increasing land productivity, however. The use of machinery does little to increase land productivity. In the Northeast where the land-labor ratio is relatively high compared to the national average, large machinery is common given that the main constraint on agriculture, especially during the peak seasons, is labor. In the South, Southeast, and North, small machinery (e.g., hand tractors) can be found. In the Northeast, machinery is used mainly for plowing, planting, and harvesting, which are labor intensive activities. In the South and East small machinery is used primarily for irrigation and transportation and other sidelines either to increase land productivity directly or to provide more opportunities for the employment of surplus labor.

Accurate figures on the irrigation of farmland are very difficult to obtain, but one of the most significant achievements since 1949 clearly has been the investment in the improvement and expansion of irrigated farmlands (see Table 2.3). The increases, however, are not well distributed. Most of the expansion has occurred in two major regions: the North and the Northeast.

The introduction of high-yield varieties also shows regional differences. The great efforts given to finding new varieties of rice since the end of 1950s have benefited the South and East regions in which rice is the traditional crop. Although some successes have been achieved in developing new strains of wheat and other crops for use in the North, the effects are less significant than those achieved in rice.

Differences in Agricultural Production Growth

The differences in the application of fertilizer and machinery, the construction of irrigation systems, and the adoption of high-yielding varieties have resulted in large differences in the growth of agricultural production among the regions. For example, in 1978 the well-developed five agricultural provinces of Jiangsu, Zhejiang, Hubei, Hunan, and Sichuan produced 63.4% of the country's commercial grain, 50% of the commercial cotton, 32% of the commercial oilseeds, and 48.7% of the commercial hogs. Although 170 commercial grain-based counties in the North, Central, South, Southeast, and Northeast regions had only 10% of the national population and 7.3% of the total arable land, they produced 15.3% of the nation's total grain output and 23.4% of the nation's commercial grain. By contrast, 241 grain-deficient counties in the Northwest and Southwest regions produced only 8.4% of the total grain crop although their share of the nation's arable land is 11.9% (Agricultural Yearbook 1980, Beijing: Agricultural Publishing House).

Agricultural production value reflects total agricultural production (see Appendix 1 for the measurement of agricultural production). The growth index of total agricultural production is shown in Table 2.4 and Figure 2.1 For the whole country, total agricultural production grew at the rate of 5% a year from 1965 to

Table 2.4 Agricultural Production Index for Different Regions
(1965 = 100)

	N.E.	N.	N.W.	C.	S.E.	S.W.	S.	National
1965	100	100	100	100	100	100	100	100
1970	137	123	105	120	119	102	119	118
1975	192	175	129	155	152	120	132	153
1976	171	162	121	146	154	114	144	147
1977	173	162	113	146	145	123	155	148
1978	198	175	120	149	162	136	165	160
1979	190	188	125	173	191	149	158	173
1980	210	205	125	164	191	166	162	180
1981	216	214	142	178	210	175	171	192
1982	226	236	156	200	237	195	196	213
1983	273	268	167	204	241	209	203	230
1984	294	303	187	227	278	230	219	257
1985	270	314	207	241	292	237	242	267
1986	276	299	205	249	321	230	300	276
Growth Rates (%)								
1965-80	4.9	4.8	1.8	3.6	4.6	3.3	3.4	4.0
1980-85	6.4	8.6	8.8	6.9	8.5	7.3	9.1	8.0
1965-85	5.3	5.7	3.5	4.4	5.6	4.3	4.8	5.0

1. Sources: China Statistical Yearbooks, various issues, Beijing: State Statistical Bureau. Agricultural Yearbooks, various issues, Beijing: Agricultural Publishing House. National Income Statistic, 1949-1985, Beijing: State Statistical Bureau.

2. Agricultural production is based on total production value. Measurements of labor and agricultural production value are shown in Appendix 1. Agricultural production value is measured in 1980 constant prices.

3. Growth rates are calculated using three-year averages.

Figure 2.1 Agricultural Production Index

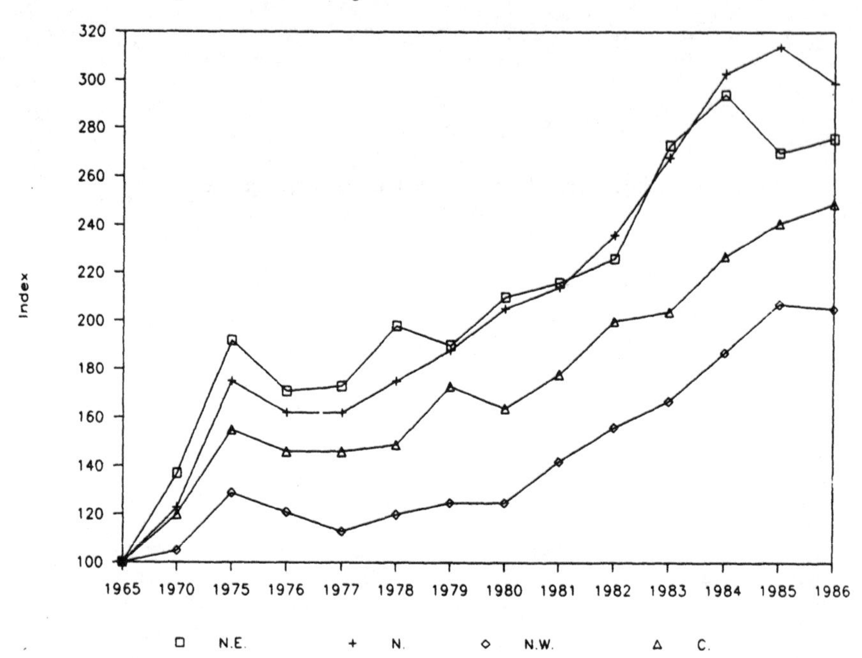

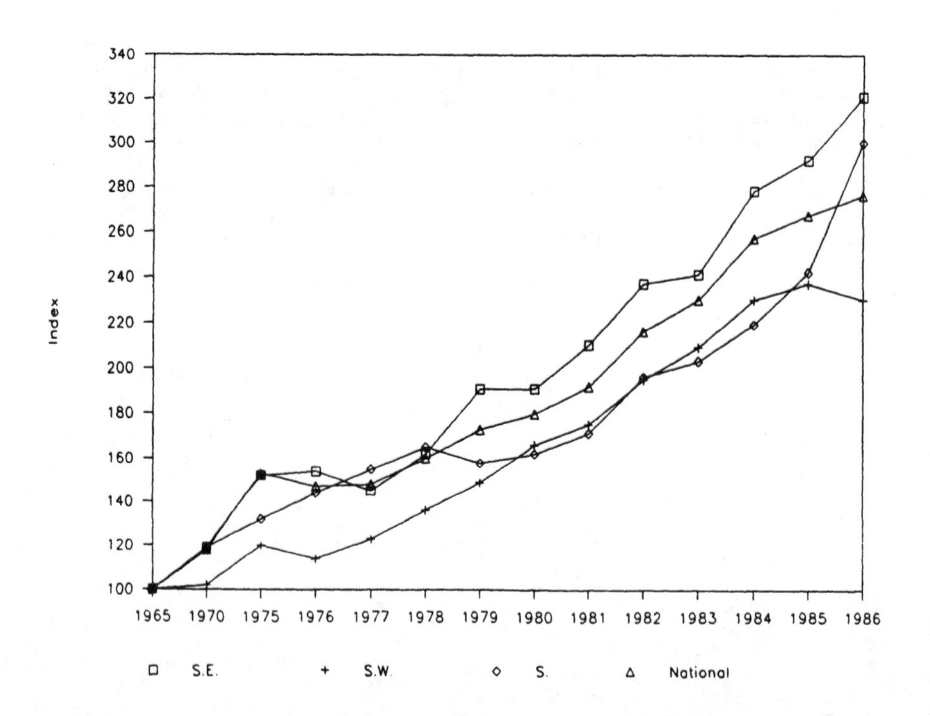

1985; up until 1980, however, the rate was 4% and then increased to 8%. This growth rate not only is the most rapid of all socialist countries but, also, is more rapid than that of most countries in the world (Hayami and Ruttan, 1971/1985, Wong and Ruttan, 1988).

Differences in regional production growth are very large. The Southeast had rapid growth rates during both 1965-1980 and 1980-1985 periods. In the South region, the growth rate was low before 1980 and high subsequently. The Northeast region had fast growth before 1980 but then the rates dropped below the national average. The Northwest region had very low growth rates before 1980 but the growth rate has accelerated in recent years. In the Central and Southwest regions, the growth rates were consistently below the national average from 1965 to 1985.

The regional growth rates of agricultural production indicate that it will not be easy for production in the low-yielding areas to attain the levels achieved in the southern and eastern coastal regions where, among other assets, water is plentiful (Perkins and Yusuf, 1984). In fact, many of the provinces in which production increased rapidly from 1965 to 1986 were high in production during the 1960s (see Appendix 2). The provinces with rapidly increasing production tend to be located on the Chinese coast or to have well-developed rural industries. Low growth is found mainly in the regions in the west, in the chronically poor interior, and in the dry provinces of Anhui and Henan on the North China Plain.

2.5 Summary

China's natural resource endowment is very poor. The limited arable land per person and variations in topography as well as weather make most of the country unfavorable for agricultural production. Thus large differences in natural resource endowment and social and economic conditions among the regions create large differences in agricultural development as well.

CHAPTER 3
ESTIMATION OF AGRICULTURAL PRODUCTION FUNCTIONS: A FRONTIER PRODUCTION FUNCTION APPROACH

3.1 Introduction

The relation between inputs and outputs is one reflection of the technology used in an economy. Estimating the production function provides such a relation. By estimating production functions for different periods and regions, the effects of technical change over time and the variations stemming from the adoption of technology can be observed.

A major purpose of this study is the comparison of partial and total factor productivities over time and among regions. Several approaches have been developed to measure total factor productivity, for example, the index number and production function approaches. The estimation of the production function has various purposes: to provide production elasticities, which are used as weights for the index number approach, and information that is essential for the production function approach. In accounting for different sources of growth over time and sources of differences among regions in total and partial productivities, approximation formulas are derived from the production function. By comparing production elasticities and factor cost shares, the degree of equilibrium can be determined also. Therefore, this chapter mainly provides information for succeeding chapters.

To start, average production function and frontier production function are identified and then the frontier production function is specified. Several production function forms are discussed before selecting the form that is the most appropriate and statistically feasible. Data and variables are explained before the main results of estimates are discussed. Finally, the results obtained here are compared with those of previous works.

3.2 Average Production Function vs. Frontier Production Function

The agricultural production function can be viewed as a physical relation between inputs and outputs in agriculture. This relation is called "technology," it embraces all the techniques used for agricultural production. It may vary over time due to technical or institutional changes.

Before the introduction of the household responsibility system in China, the basic production unit was the production team; the team leader made decisions on the use of available inputs and outputs to fulfill government targets primarily and secondly, to maximize profits. Currently, under the household responsibility system, farmers make their decisions on inputs and outputs on the basis of maximizing profits after fulfilling the quotas contracted with the government. In the past, a physical relation existed between the inputs and outputs in production teams. The relation between inputs and outputs still exists on farms under the household responsibility system. We assume that all teams in the past and farmers today have potential access to the same technology and choose different points on the technology, depending on the relative scarcity of resource endowment. However, because of inefficiency, realized output for a particular farm may not reach its potential output.

For the sake of simplicity, let us consider two inputs (e.g., labor and land) and one output case, as shown in Figure 3.1, The curves pc_1 and pc_2 represent production isoquants for regions 1 and 2, respectively. If production were perfectly efficient technically, region 1's output would be situated at point A, depending on its factor ratio, whereas region 2 would produce at point B, depending on its factor ratio. However, due to the inefficiencies, the input combinations for the same outputs may be A^* for region 1 and B^* for region 2. Therefore, the relative efficiency of production is OA/OA^* for region 1, and OB/OB^* for region 2.

The traditional estimation of production function assumes that every firm is technically efficient, resulting in the average production function. This assumption is unrealistic. Given the available technology, a firm may not reach potential output. The

Figure 3.1 Technical Inefficiency in Production

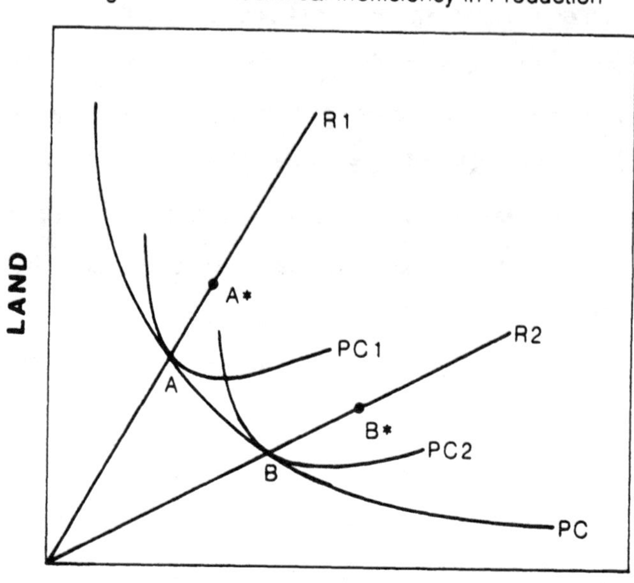

differences between realized output and potential output (or frontier) are caused by production inefficiency. However, through learning by doing, diffusion of new technological knowledge, improved managerial practice, and institutional change, the distance between realized and potential output may be decreased. For example, before the institutional change in Chinese agriculture in 1979, production was organized by production teams or state farms. The farmers' income was not totally dependent on the efforts they devoted to production, then farmers might not make a full effort to produce and labor was used inefficiently. Under such circumstances, disincentives prevailed in agricultural production.

After 1979, farmers worked for themselves and they worked hard and more efficiently. Therefore, given the same inputs, more output has been achieved. If we ignore efficiency improvement by attributing it to technical change, then we will overestimate the effects of technical change on productivity growth and production growth.

Since our purpose is to compare the productivity growth pattern and the sources of productivity growth among regions, we will use panel data. However, as Lau and Yotopoulous (1987) noted, using panel data leads to some problems. The first problem is the possible existence of interregional differences in the definitions, measurements, and quality of the outputs and inputs. For example, the composition of outputs and inputs may be different across regions. The second problem is the existence of interregional differences in the basic environment, which include climate, topography and infrastructure (discussed in Chapter 2). They may lead to differences in the levels of technical efficiency, that is, in the ability to produce outputs from given quantities of inputs.

Therefore the production function we estimate should reflect both technical change and efficiency change. The frontier production function approach makes it possible to study these two factors together.

3.3 Specification of Frontier Production Function and Choice of Production Functional Forms

The specification of frontier production function is discussed first and then the choice of production function form is presented.

Specification of Frontier Production Function

Initiated by Farrell in 1957, the frontier production function approach has been well developed recently. (A good summary of frontier approaches was made by Lovell and Schmidt in 1988.) Various methods for measuring and computing frontier production functions and efficiency have been expanded. The main approaches include the pure programming approach by Farrell (1957), and Charnes, Cooper, and Fare (1984); the modified programming approach by Farrell (1957), Aigner and Chu (1977), Forsund and Jansen (1977), and Forsund and Hjalmarsson (1979), among others; the deterministic statistical frontier approach by Afriat (1972), Richmond (1974), and Greene (1980), among others; and the stochastic frontier approach by Aigner, Lovell, and Schmidt (1977), and Meeusen and van den Broeck (1977), among others.

Pitt and Lee (1981) indicated that the programming approach and the deterministic frontier approach do not allow for random shocks that are outside a firm's control in the production process. As a consequence, a few extreme measured observations determine the frontier and exaggerate the maximum possible output given the inputs. In this study, the stochastic frontier approach is employed to avoid random shock as the source of inefficiency.

Consider the following production function model:

$$Y_{it}=f(x_{it},b)e^{v_{it}}e^{u_{it}}, \tag{3.1}$$

where i denotes the ith firm or region, and t denotes time

t. Y_{it} is output, x_{it} is 1×k rows of inputs, $f(x_{it},b)$ is theoretical potential output, v_{it} is a stochastic variable which represents uncontrolled random shocks in the production, and u_{it} is one-sided distribution, $u \leq 0$, which represents technical inefficiency. $f(x_{it},b)e^{vit}$ is the stochastic frontier, since v_{it} consists of random factors outside the firm's control. The non-positive disturbance u reflects the fact that output must lie on or below its frontier $f(x_{it},b)e^{vit}$, since e^{uit} has a value between zero and one. We assume that for $t \neq t'$, $E(u_{it}u_{it}') = 0$ for all i and $E(u_{it}u_{it}') = 0$ for all $i \neq j$. In this specification, the firm's inefficiency may change over time since the firm may improve its efficiency through learning by doing. We also assume u is truncated normal with variance σ_u^2 and v is normal with mean zero and variance σ_v^2.

The efficiency for a firm or region i at time t is defined as:

$$\frac{Y_{it}}{f(x_{it},b)e^{vit}}.$$

Based on the conditional distribution of u_{it} given the distribution $v_{it} + u_{it}$, we can measure the efficiency of a specific firm or region at a given time. Following Kalirajan and Flinn (1983), The efficiency can be measured as

$$E\left\{\exp\left(\frac{u_{it}}{u^{it} + v^{it}}\right)\right\} = \exp\left[-\left(\frac{\sigma_u \sigma_v}{\sigma}\right)\left(\frac{f(.)}{1-F(.)} - \frac{e_{it}}{\sigma}\frac{\sqrt{\lambda}}{1-\lambda}\right)\right] \quad (3.2)$$

where $e_{it} = v_{it} + u_{it}$, σ is standard error of e_{it}, $\lambda = (\sigma_u^2)/(\sigma^2)$, and f(.) and F(.) are the values of the standard normal density function, and the standard normal distribution function is evaluated at

$$\frac{e_{it} \sqrt{\lambda}}{\sigma \quad 1-\lambda.}$$

The average efficiency for different times can also be calculated. Following Lee and Tyler (1978), with the truncated normal distribution, the mean efficiency is measured by

$$E\left(\exp(u)\right) = 2\times\exp(\sigma_u^2/ 2)(1-h(\sigma_u)); \qquad (3.3)$$

where h is the standard normal cumulative density function.

Choice of Production Function Form

Production function forms should satisfy the following criteria: (1) The functional forms should contain no more parameters than are necessary to agree with the maintained hypotheses. (2) The parameters should have intrinsic and intuitive economic interpretations, and a clear functional structure. (3) The trade off between the computational requirements of a functional form and the roughness of empirical analysis should be weighed carefully in the choice of a model. (4) The chosen functional form should be well behaved, and should be consistent with the maintained hypotheses, such as positive marginal products or convexity, within the range of observed data. (5) The functional form should be compatible with the maintained hypotheses outside the range of observed data (Fuss, McFadden, and Mundalak, 1978).

The Cobb-Douglas form receives the widest application because it is simple and has clear economic implications. The Cobb-Douglas production function, however, implies constant production elasticities and for China it is unrealistic to assume such constancy. Because land is scarce and labor abundant, efforts at technological improvements concentrate on saving land and increasing output per unit of limited land areas. These aims are achieved primarily through the construction of large-scale water control projects and capital construction of farmland, the development of fertilizer-responsive, high-yielding varieties, and other techniques. Such bias in technical change also is reflected in the decreased shares of land and labor, and in the corresponding increased shares of current inputs in the total cost of agricultural production (see Chapter 4).

Recent years have seen the introduction of more general production functional forms such as the translog, generalized quadratic, generalized Leontief, and Box-Cox. The more general forms have more variables and lesser restrictions but they require more data, which may lead to some econometric estimation problems, such as multicollinearity. This problem occurs especially in Chinese agriculture where fertilizer, machinery, and other current inputs have similar time trends.

Consider a production process that uses n inputs to produce one output which is represented by the production function,

$$Y = f(x_1,\ldots,\ldots x_n,\ T), \qquad (3.4)$$

where Y is output, x_i is i^{th} input and T is used to catch technical progress (Time trend). The translog production function can be used to represent the production function (3.4).

$$\ln(Y) = a_0 + a_t t + \sum_i a_i \ln(x_i) + \sum_i a_{it} \ln(x_i) \times t + \sum\sum_{ij} a_{ij} \ln(x_i) \times \ln(x_j) + a_{tt} t^2. \quad (3.5)$$

If the production function is linear-homogeneous in inputs, then the following constraints are imposed:

$$\sum_i a_i = 1;$$
$$\sum_i a_{ij} = 0; \text{ for all } j;$$
$$\sum_i a_{it} = 0.$$

Consider a restriction that all inputs are separable from each other, but each input cannot be separated from technical progress:

$$Y = f\{g_1(x_1, T), g_2(x_2, T) \ldots g_n(x_n, T)\}. \quad (3.6)$$

The theoretical background of this form is the fact that every input changes over time whereas the effects among inputs are indirect through time. (This is a special case of (3.4).) Then the following production function form can be used to represent (3.6):

$$\ln(Y) =$$
$$a_0 + a_t t + \sum_i a_i \ln(x_i) + \sum_i a_{it} \ln(x_{it}) \times t + a_{tt} t^2. \quad (3.7)$$

The constraints for the linear-homogeneous production function of (3.7) are as follow:

$$\sum_i a_i = 1;$$
$$\sum_i a_{it} = 0.$$

If we consider all inputs and time to be separable, then the production function can be expressed as

$$Y = f\{g_1(x_1), \ldots, \ldots g_n(x_n), T\}. \quad (3.8)$$

The Cobb-Douglas production function can be used to represent (3.8).

$$\ln(Y) = a_0 + \Sigma_i a_i \ln(x_i) + a_{it} t. \qquad (3.9)$$

The restrictions for a linear-homogeneous production function of (3.9) is as follows:

$$\Sigma_i a_i = 1.$$

Here we examined the candidates of production forms. Due to the serious multicollinearity problem of translog form, the estimated results are not significant. It has been argued that the constant production elasticities implied in the Cobb-Douglas production function are not a critical limitation for a short period of time, but this assumption is unrealistic for a long period. Hence, we use the Cobb-Douglas production function form for cross-section data to estimate the production function for each year. The production function form (3.7) is employed to estimate the whole period production function in order to avoid the constancy of production elasticities over time and to escape the serious multicollinearity problems of the translog form. (Of course, we also can use time slope dummies, but they have more variables and the estimated results are affected greatly by each year's random shock.)

3.4 Variable and Data Explanation

Cross-section and time-series data are explained in this section.

After 1980, China began to publish more annual data systematically. Some of the publications are the yearbooks produced by the Statistic Publishing House; State Statistical Bureau; and Agricultural Publishing House of Ministry of Agriculture, Livestock Husbandry, and Fisheries. Recently, some pre-1980 data also have been released. Hence national time series data from 1952 to 1986 have become available although the provincial data before 1978 are incomplete. Some statistical techniques have been used to interpolate the missing data (see Appendix 1).

Agricultural Output. Gross agricultural production value is used as the aggregate total output. It is aggregated by each product times its price (1980 constant prices). The sub-aggregates are (a) crop production, (b) forestry, (c) animal husbandry, (d) sideline industries, and (e) fisheries. Rural industry is excluded from agricultural production because it does not have agricultural characteristics and is less likely to have the same production function as crop production, forestry, animal husbandry, fisheries, and sideline industries. However, in Tang's (1984) study, gross agricultural value includes the value of rural industry (see Appendix 1 for more detail).

Land Input. The sum of sown areas and pasture is used to measure land input in China because the arable land data are extremely inaccurate. Sown land plus pasture also indicates quality more accurately. Pasture areas are calculated in sown land area equivalence for output value, that is one unit of pasture equals .0124 of a unit of sown land (in 1985).

Labor Input. In this study, we measure labor input in agriculture in stock terms. It is measured in terms of numbers of person at the end of the year (see Appendix 1 for more detail). The assumption is that the ratio of labor to population has the same growth rate in every region. The labor of rural industry is excluded from total agricultural labor. However, in past studies, the labor input was measured in terms of all labor in the rural areas (Wong, 1986; Tang, 1983).

Machinery Input. Machinery input for the purpose of this study is measured in terms of total horsepower at the end of each year. In some years, no regional statistics on horsepower are available, only data on tractors for each province and national total horsepower. Thus, to indicate shares of horsepower for each region, shares of the national total of tractors are used to interpolate regional horsepower.

Draft Animal Input. Draft animals are measured at the end of the year which are used for agricultural activities and rural transportation.

Chemical Fertilizer Input. This input is measured by pure nutrient. The pure nutrient of fertilizer input before 1978 was estimated from gross chemical fertilizer (see Appendix 1).

Manurial Fertilizer Input. Manurial fertilizer always has been very important, both earlier and currently. In recent years, chemical and mineral fertilizers have become increasingly popular although the traditional manurial resources continue to be the main source of nutrient materials. Manurial resources include animal, human, and crop wastes; green manures; and water plants. Here manurial fertilizer is measured from the agricultural population (i.e., human waste) and numbers of domestic animals.

3.5 Results of Estimation of Production Functions

By using different specifications of production functions, the frontier production functions for different years and for the whole period of 1965 to 1986 are estimated in this section. (The production functions for different periods and the production function for the 1965-86 period with fixed and random effects models are shown in Appendix 4.)

Frontier Production Functions for Different Years

Using the Cobb-Douglas form, the estimates of frontier production functions for different years are presented in Table 3.1.

Except for irrigated areas and machinery input, most estimates are statistically significant. The steady increase of the constant terms suggests that the neutral technical change shifts the production isoquant to the origin steadily. The coefficients of labor and land are rather large all the time although they are decreasing, which indicates that traditional inputs are still significant in agricultural production. The increasing coefficients of chemical fertilizer input imply that in China, fertilizer is used as a substitute for scarce land.

Table 3.1 Estimates of Frontier Production Functions for Different Years

Year	Constant	x_1	x_2	x_3	x_4	x_5	x_6	x_7	$E(e^u)$
65	-2.71	.290	.352	.213	.032	.153	-.018	.039	.745
	(-1.06)	(2.19)	(1.35)	(5.28)	(.544)	(.227)	(-.865)	(1.02)	
70	-2.87	.336	.319	.159	.059	.220	-.009	.027	.788
	(-3.65)	(1.75)	(1.68)	(1.75)	(.621)	(1.88)	(-.069)	(.167)	
75	-3.41	.247	.592	.236	.112	.084	-.143	-.051	.789
	(-5.91)	(.27)	(3.23)	(1.60)	(.70)	(.41)	(-1.75)	(-.29)	
76	-2.21	.433	.021	.128	.268	.311	-.062	-.003	.785
	(-3.02)	(2.34)	(.179)	(1.13)	(1.99)	(2.05)	(-.57)	(-.018)	
77	-2.87	.318	.368	.230	.173	.171	-.064	-.089	.782
	(-5.55)	(1.24)	(2.27)	(2.26)	(1.77)	(1.18)	(-.40)	(-.38)	
78	-2.77	.195	.457	.236	.297	.244	-.159	-.172	.816
	(-6.09)	(1.88)	(4.32)	(1.91)	(3.37)	(2.91)	(-2.38)	(-.3.37)	
79	-3.53	.232	.583	.148	.260	.206	-.202	-.139	.799
	(-3.80)	(1.06)	(3.18)	(1.20)	(1.28)	(1.65)	(-1.51)	(-.59)	
80	-2.26	.189	.520	.260	.058	.176	-.124	-.105	.828
	(-5.82)	(2.85)	(6.99)	(6.17)	(.85)	(2.93)	(-1.59)	(-.2.23)	
81	-2.13	.068	.460	.288	.037	.295	-.143	.011	.829
	(-2.86)	(.216)	(2.88)	(2.23)	(.21)	(3.49)	(-.98)	(.04)	
82	-2.10	.164	.523	.315	.050	.149	-.153	-.083	.905
	(-2.87)	(1.38)	(3.34)	(3.40)	(.334)	(1.17)	(-3.08)	(-.609)	
83	-2.00	-.079	.401	.208	.127	.394	-.184	-.055	.893
	(-3.36)	(-1.08)	(.54)	(1.17)	(2.24)	(2.13)	(-.57)	(-.41)	
84	-1.37	.105	.320	.456	.150	.121	-.092	-.058	.886
	(-1.70)	(1.09)	(2.26)	(5.00)	(.98)	(1.78)	(-2.02)	(-.26)	
85	-1.19	.039	.281	.425	.113	.254	-.142	.037	.888
	(-1.13)	(.27)	(1.99)	(3.35)	(1.03)	(2.89)	(-1.91)	(.35)	
86	-1.36	.324	.105	.192	.335	.116	-.089	-.016	.783
	(-.93)	(1.31)	(.52)	(1.16)	(1.84)	(.73)	(-.44)	(-.08)	

1 x_1: labor; x_2: land; x_3: chemical fertilizer; x_4: machinery; x_5: manurial fertilizer; x_6: draft animal; x_7: irrigated areas.

2. Observation numbers for each regression are 29.

3. The numbers in parenthesis are t test values.

4. Derivation of log-likelihood function and econometric estimation procedure is presented in Appendix 5.

The trend of machinery coefficients, although it is increasing, is not very obvious.

The last column of the table shows the average efficiency for each year (calculated by using (3.3)). The efficiency of production was stagnant before 1979 but improved greatly after 1980 as a result of institutional change.

Frontier Production Functions Using Pooled Data

In order to use all the information of panel data (i.e., pooling the cross-section and time-series data), we estimate the production functions for the entire period of 1965-1986.

The common Cobb-Douglas form is used for regression 1 in Table 3.2. The variables are labor, land, draft animals, manurial fertilizer, and irrigation (the representative of rural infrastructure); they catch the effects of conventional inputs on the production. Time trend measures the technical changes occurring over time. Except for the variables of machinery and irrigation, the coefficients are very significant. The negative coefficients of draft animals are not the case in reality. The sum of production elasticities of traditional inputs is more than .75 (except for draft animals), which implies that traditional inputs still dominate China's agricultural production. The chemical fertilizer input has an important role in production whereas the contribution of machinery input to production is insignificant. The significance and positivity of the

Table 3.2 Estimates of Agricultural Frontier Production Functions for the Whole Period of 1965-1986, Using Panel Data

Regression Number	R1 (Average)	R2 (Frontier)	R3 (Frontier)	R4 (Average)	R5 (Frontier)
Constant	-2.807	-2.703	-2.81	-3.19	-3.255
	(-10.72)	(-11.27)	(-5.23)	(-6.13)	(-6.72)
x_1	.278	.266	.420	.417	.358
	(7.19)	(6.14)	(5.16)	(4.66)	(4.14)
x_2	.356	.379	.243	.331	.4004
	(7.88)	(9.39)	(2.40)	(3.99)	(5.91)
x_3	.235	.236	.140	.089	.157
	(8.71)	(9.29)	(2.704)	(1.66)	(2.94)
x_4	.055	.051	.078	.123	.003
	(1.77)	(1.82)	(1.39)	(2.52)	(.030)
x_5	.185	.178	.227	.266	.180
	(5.30)	(5.67)	(2.99)	(3.27)	(2.51)
x_6	-.132	-.133	.002	-.026	
	(-5.13)	(-4.94)	(.037)	(-.301)	
x_7	.059	.055	.009	-.037	.0124
	(1.81)	(1.66)	(.145)	(-.537)	(.197)
x_t	.0123	.0125	.0014	.042	.0709
	(2.41)	(2.17)	(.364)	(.980)	(1.859)
x_{t1}			-.0097	-.0109	-.011
			(-1.822)	(-1.79)	(-1.79)
x_{t2}			-.0024	-.0065	-.016
			(-.368)	(-1.20)	(-3.65)
x_{t3}			.0068	.0087	.0077
			(1.83)	(2.406)	(2.15)
x_{t4}			.008	.0083	.0195
			(1.93)	(2.08)	(3.17)

(Continued)

Table 3.2 (Continued)

Regression Number	R1 (Average)	R2 (Frontier)	R3 (Frontier)	R4 (Average)	R5 (Frontier)
x_{t5}			-.00006	-.0014	-.0038
			(-.013)	(-.273)	(-.83)
x_{t6}			-.006	-.0041	
			(-1.51)	(-.725)	
x_{t7}			-.0003	.0006	.0014
			(-.064)	(.118)	(.261)
x_{tt}			.00147	.0013	.00023
			(2.23)	(2.30)	(.350)
Average Efficiency		.871	.846		.858
Observations	406	406	406	406	406

1. Numbers in parentheses are t test values.

2. x_1: labor; x_2: land; x_3: chemical fertilizer; x_4: machinery; x_5: manurial fertilizer; x_6: draft animals; x_7: irrigated areas; x_{t1}: cross term of labor and time trend; x_{t2}: cross term of land and time trend; ......; x_{t7}: cross term of irrigated areas and time trends.

3. Average: Average production function ; Frontier: Frontier production function.

4. Machinery input in Regression 5 is the sum of machinery horsepower and number of heads of draft animals.

5. Ordinary least squares is used for the average production function estimation and maximum likelihood technique is used for the frontier production function estimation.

6. Production function form (3.7) is used for regressions 3, 4, and 5.

7. The production elasticity for input i in this function form is $\partial \ln Y / \partial \ln x_i = a_i + a_{it}t$. Therefore, if $a_{it} > 0$, production elasticity of input i is increasing, implying input i using; if $a_{it} < 0$, production elasticity of input i is decreasing, implying input i saving.

time trend coefficient strongly suggest the existence of technical change in China's agriculture.

Regression 2 uses the frontier production function approach. The results are very much the same as those for regression 1.

Regression 3 uses functional form (3.7); it includes the same input variables as those for regressions 1 and 2. In addition, the cross term of each input and time trend is used to catch the relative changes of each input in total input over time. Like regressions 1 and 2, the results are significant except for the coefficients of draft animals and irrigation, considering the crudeness of data. Labor, land, draft animals, and manurial fertilizer are decreasing in importance in production whereas the production elasticities of chemical fertilizer, manurial fertilizer, and machinery have been increasing over time. The coefficients of both t and t square are positive and significant.

The average production function is estimated in regression 4 for comparison with regression 3, using functional form (3.4). The results are similar to those of regression 3.

In regression 5, machinery and draft animals are aggregated as one variable in order to escape the negative coefficient of draft animals. The substitution elasticity of machinery for draft animals is assumed to be perfect here. The frontier approach is employed in the estimation and the results are similar to those of regression 3.

Technical Efficiency for Each Region

Using (3.4) and the results of regression 3 in Table 3.2, the efficiency for each region was calculated and is shown in Table 3.3.

The Northeast region had the highest technical efficiency during most of the periods (1965 to 1985). In fact, in the 1960s and 1970s land/labor ratios were very favorable, and the efficiency loss in the Northeast from the existence of surplus labor in agriculture was not as serious as in other regions. However, as national average efficiency increased following the institutional change at the end of the 1970s and beginning of 1980s, the efficiency of production in the Northeast decreased.

The North region had the lowest efficiency in agricultural production during most of the 1965-1985 period. The provinces of Gansu, Shaanxi, and Henan provinces in the North region are among the poorest provinces in China. The insufficiency of water and the large surplus of labor have resulted in low efficiency in agricultural production.

Efficiency was relatively high in the Northwest in 1965 probably due to the favorable land/labor ratio. However, the efficiency decreased and the region became one of the least efficient in 1985.

The Central region had low production efficiency during the 1960s but it has been improved greatly since 1970. Today, the region is one of the most efficient regions in the country.

Unlike growth in output, the production efficiency

Table 3.3 Technical Efficiency for Each Region (Selected Years)

Year	N.E	N.	N.W.	C.	S.E.	S.W.	S.	National
1965	.933	.709	.869	.796	.832	.885	.685	.745
1970	.900	.659	.849	.847	.873	.736	.725	.788
1975	.931	.777	.852	.859	.805	.778	.734	.789
1980	.934	.747	.739	.894	.794	.861	.936	.828
1985	.916	.818	.829	.952	.877	.927	.954	.888

in the Southeast has changed only slightly. In 1985, the production efficiency in the region was one of the lowest (next to the North).

In the Southwest the relatively high efficiency in agricultural production offset the low total production growth. In fact, the efficiency in 1985 was the highest nationally.

The technical efficiency in the South has increased greatly since 1965. It changed from one of the least efficient regions to one of the most efficient. The improvement of efficiency is crucial to the rapid growth of the region's total production.

3.6 Comparison with Previous Estimates

The results of the estimation of frontier production functions and average efficiency are compared here with the previous studies. By the comparison, we find that our results do not differ significantly from those of other investigators.

Comparison of Production Function Coefficients

Inasmuch as the Cobb-Douglas production function form is employed in most of the previous studies, we compare only the results of regressions 1 and 2 in Table 3.2 with those of others. Comparisons are presented in Table 3.4.

Wong's estimation of the production functions for nine socialist countries resulted in very significant findings. I do not consider his estimates to be inconsistent with those of the present study except for livestock and fertilizer. His exclusion of manurial fertilizer from inputs cause the inconsistency of fertilizer coefficient with the results of this study. Hayami and Ruttan (1971/1985) estimated production functions for less developed countries. They showed that labor production elasticity for developing countries seems to be larger than that in China. The negative production elasticity of land is not the case in practice. Lau and Yotopoulous (1987) employed a translog form, based on the first differenced data, using the same data set as did Hayami and Ruttan, and achieved results similar to that presented here except for fertilizer and livestock.

Comparison of Average Productive Efficiency

The estimation of average productive efficiency presented here can be checked with those of earlier studies (Table 3.5). The mean efficiency of China's agriculture ranges

Table 3.4 Comparison of Agricultural Production Elasticities

	This Study R1 (OLS)[c]	R2 (Frontier)	Wong (1986) (MR)[d]	H-R[a] (1985) (OLS)	L-Y[b] (1987) (FD)[e]
Observations	406	406	272	43	43
Labor	.278	.266	.223	.562	.396
	(7.19)	(6.14)	(24.78)	(5.45)	(5.64)
Land	.356	.379	.143	-.065	.403
	(7.88)	(9.39)	(28.60)	(-1.03)	(3.28)
Fertilizer	.426[f]	.414[f]	.177	.089	.058
	(7.01)	(7.53)	(19.7)	(3.74)	(2.46)
Machinery	.055	.051	.122	.136	.109
	(1.77)	(1.82)	(8.13)	(2.57)	(4.68)
Livestock	-.132	-.133	.233	.318	.143
	(-5.13)	(-4.95)	(.176)	(3.74)	(2.63)
Irrigation	.059	.055			
	(1.81)	(1.66)			
Time	.0123	.0125			
	(2.41)	(2.17)			
Sum of Coef	1.042	1.032	.898	1.04	1.109

Notes: a. H-R: Hayami and Ruttan (1985).
b. L-Y: Lau and Yutopoulous (1987)
c. OLS: ordinary least squares.
d. MR: Mixed Regression.
e. FD: First Difference.
f. The coefficient of fertilizer for R1 and R2 is the sum of chemical and manurial fertilizer coefficients.

from 84.6% with regression 4 to 87.1% with regression 2. These are higher than the 67.7% average efficiency found in the Indonesian weaving industry (Pitt and Lee, 1981), the 71.73% found for Canadian farms (Turvey and DeBore, 1988) and the 69.0% for the Australian dairy industry (Battese and Coelli, 1988), but are somewhat lower than the 92.9% for the Soviet cotton refining enterprises (Danilin et al., 1985) and the 89.75% for Indian farms (Huang and Bagi, 1984).

Table 3.5 Comparison of Average Efficiency Estimation

Investigators		Estimation Approach	Efficiency
	R2	Stochastic Frontier	.871
Fan (1989)	R3	Stochastic Frontier	.846
	R5	Stochastic Frontier	.858
Pitt and Lee (1981)		Stochastic Frontier	.6772
Huang and Bagi (1984)		Stochastic Frontier	.8975
Danilin et al. (1985)		Stochastic Frontier	.929
Turvey and DeBore (1988)		Covariance Analysis	.7173
Battese and Coelli (1988)		Stochastic Frontier	.690

CHAPTER 4
COMPARISON OF PARTIAL PRODUCTIVITY AND TOTAL FACTOR PRODUCTIVITY GROWTH AMONG REGIONS

4.1 Introduction

In Chapter 2 the regional characteristics of resource endowment and the social and natural conditions that have determined the different growth paths of partial productivities and total factor productivity were discussed. The physical relation between inputs and outputs was estimated in Chapter 3. Based on the measurement of partial productivity and total factor productivity, a comparison of productivity changes over time among regions is conducted and the resulting interrelation of various productivity changes is analyzed in this Chapter.

The comparisons for 1965 and 1986 reveal not only that interregional differences in the partial productivity ratios were very large but, also, that they increased over the period.

By comparing the production elasticities and factor shares, the disequilibrium in China's agricultural production is examined. Both production elasticities and factor shares are used to calculate the total factor productivity index for the comparison.

4.2 Regional Comparison of Partial Productivity Growth

Partial productivities usually consist of labor productivity, land productivity and capital productivity. Capital productivity is not widely used in agriculture, hence labor and land productivity growth is compared in this section.

Labor Productivity Growth

Labor productivity is often used to measure economic progress. Agricultural labor productivity is used to measure the population supported by each agricultural labor.

It is useful to partition labor productivity into two components, the land/labor ratio and land productivity.

$$Y/L = (A/L)(Y/A) \qquad (4.1)$$

where Y is output, L is labor input, and A is land input.

The labor productivity for different regions is shown in Table 4.1 and labor productivity index in Figure 4.1 (see Appendix 2 for the labor productivity for the different provinces). In China, labor productivity is low compared to that of other developing or socialist countries because of the extremely low land/labor ratio. However, from 1965 to 1986 the growth rate of China's labor productivity was rather high despite the decrease of land areas per labor unit. Since land input already

Table 4.1 Labor Productivity for Different Regions
(1980's Yuan)

	N.E.	N.	N.W.	C.	S.E.	S.W.	S.	National
1965	785.3	340.7	823.0	503.3	436.2	390.2	449.7	439.60
1970	776.0	356.9	722.3	503.2	438.9	325.0	444.4	432.15
1975	998.0	478.3	811.5	589.3	506.7	341.0	443.2	510.14
1976	889.0	445.9	755.8	557.8	515.0	321.0	482.3	490.62
1977	905.1	450.1	703.4	561.5	488.3	346.5	524.0	497.12
1978	1169.6	551.3	796.9	649.5	617.4	438.6	626.6	609.36
1979	1463.3	613.3	803.3	764.1	738.9	474.6	599.7	673.27
1980	1608.7	614.8	887.1	682.8	677.7	494.8	576.2	659.03
1981	1620.3	629.0	979.8	723.3	815.0	503.3	597.1	697.70
1982	1665.8	676.8	1052.4	793.3	824.0	544.8	672.9	746.27
1983	1859.1	764.4	1080.9	802.3	827.3	572.9	687.2	793.45
1984	2009.5	868.5	1205.2	871.8	959.9	627.0	740.2	883.86
1985	1879.9	941.8	1305.9	966.8	1118.3	650.7	846.2	959.49
1986	1870.3	897.5	1338.2	1007.6	1241.5	620.2	1040.8	989.24
Annual Growth Rate %								
1965 to 79	4.90	4.62	-.19	3.26	4.14	1.52	2.24	3.33
1980 to 86	2.54	6.51	7.09	6.70	10.61	3.84	10.36	7.00
1965 to 86	4.20	4.70	2.30	3.36	5.10	2.23	4.08	3.94

1. Labor productivity is measured by total agricultural production divided by total labor input. Total agricultural production value is measured in terms of 1980's yuan. The data for the years after 1978 are taken from China's Statistical Yearbook, 1981 to 1987, Beijing: State Statistical Bureau. The data prior to 1978 are taken from National Income Statistics, 1952 to 1985. Beijing: State Statistical Bureau. Because total agricultural production value provided for each province before 1978 includes the production value of its rural industry, I have subtracted the estimated rural industrial value is from the agricultural total production value. The share of each region's rural industrial value in the national total in 1979 is used to interpolate the rural industrial value before 1979 for each province.

2. Agricultural labor after 1979 is taken from China's Rural Statistical Yearbook, 1985 to 1987. Labor inputs before 1979 are estimated from agricultural population, and agricultural population is taken from National Agricultural Statistics for 30 years (1949 to 1979), Beijing: State Statistical Bureau, March, 1980.

Figure 4.1 Labor Productivity Index

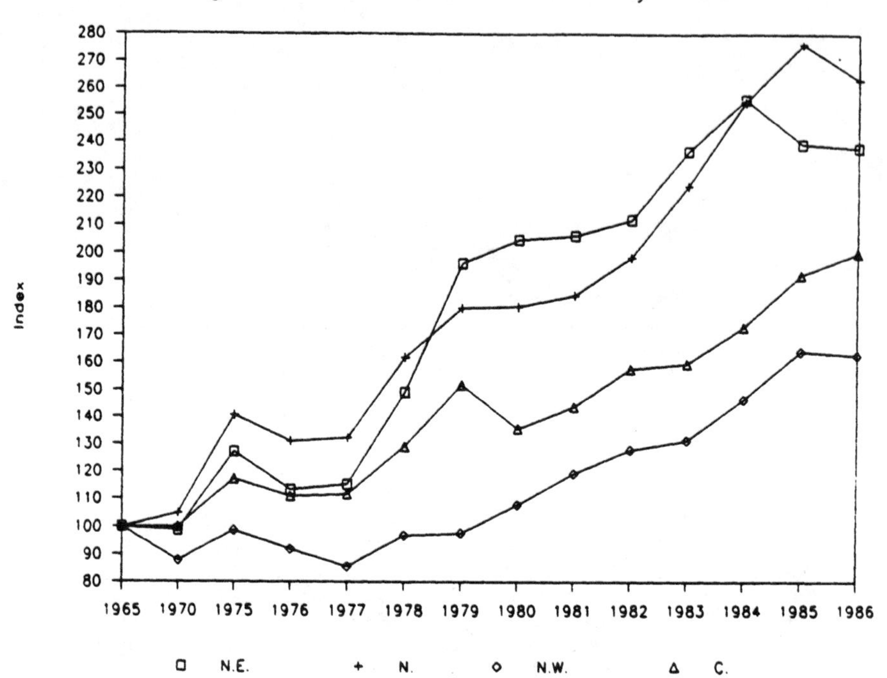

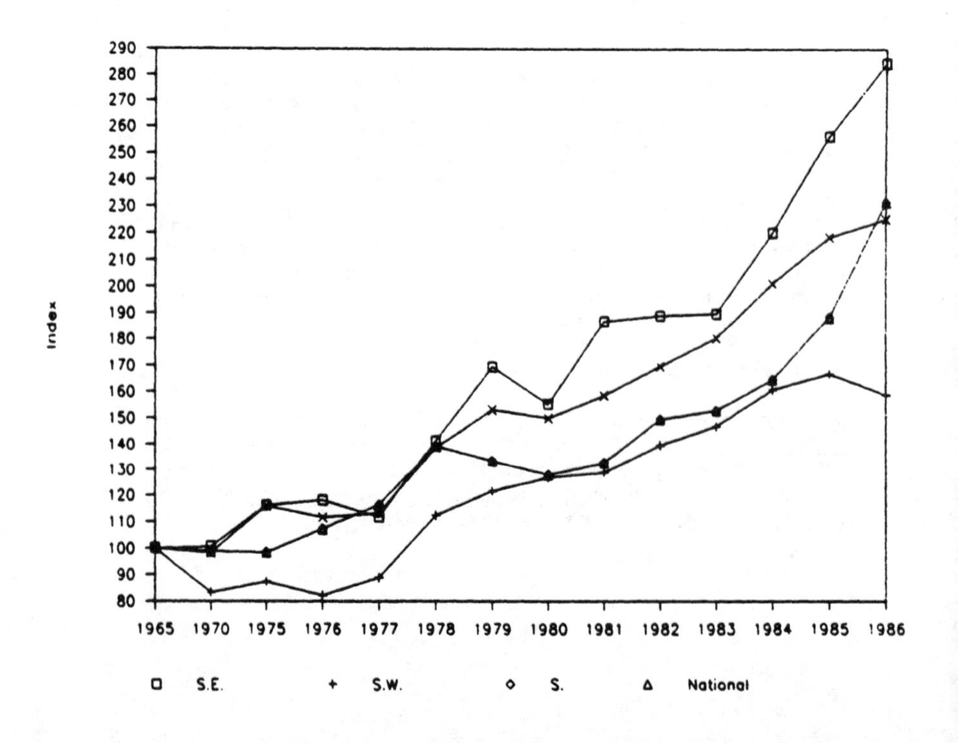

included land on the margin, the increase of land productivity through labor-intensive farming patterns has mainly contributed to labor productivity.

Interregional differences in labor productivity are large. In general, the regions with high land-labor ratios have higher labor productivity. The Northwest region was considered the most backward region in agricultural production. However, labor productivity in this region was the highest among all the regions in 1965 and is still one of the highest today. In 1965, Yuan per labor unit in the Northwest region was 2.4 times that of the North region (822.97 to 340.7). Labor productivity in the Northeast (1870.3 yuan per labor unit) was three times that in the Southwest (620.2 yuan per labor unit) in 1986. The differences have been increased. In the Northwest region, land productivity growth was slower than land-labor ratio change as a result of fast population growth; labor productivity actually showed negative growth from 1965 to 1979. The Northeast passed the Northwest as the region with the highest labor productivity after 1970.

Every region had higher growth rates after 1979 except the Northeast. The institutional change accelerated the growth after 1979 which had two positive effects on labor productivity: First, farmers had incentives to increase total output by allocating resources (especially labor) more efficiently. Second, new institutional reforms encouraged off-farm activities, such as rural industry and rural services; they absorbed considerable numbers of surplus labor from agriculture and

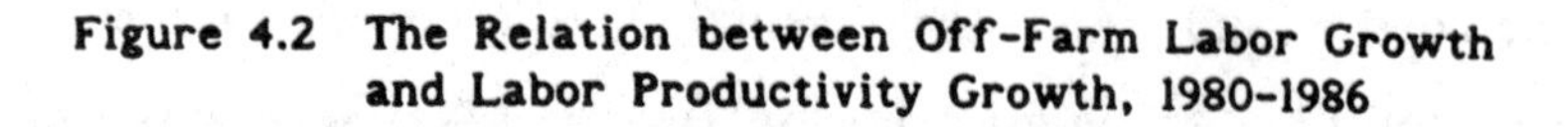

Figure 4.2 The Relation between Off-Farm Labor Growth and Labor Productivity Growth, 1980-1986

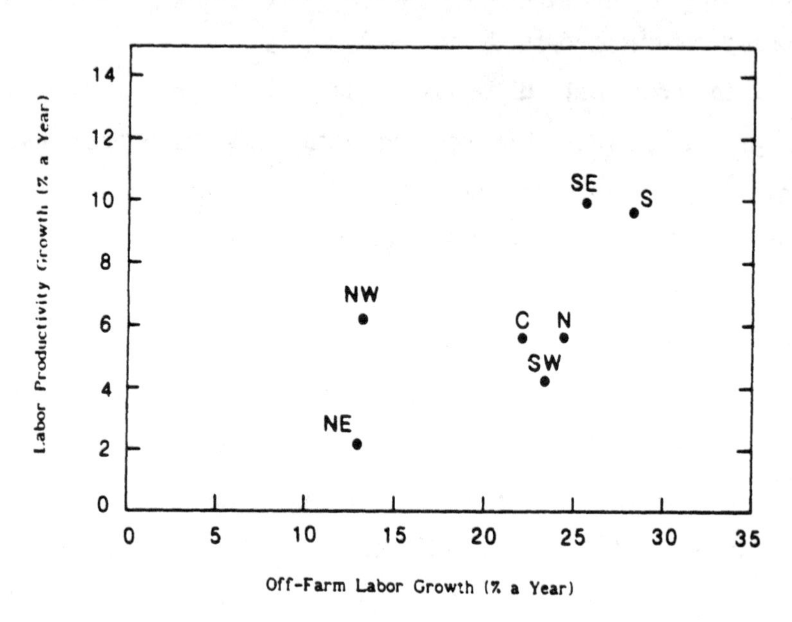

thereby increased the land-labor ratio. Thus, labor productivity growth after 1979 has positively correlated with the growth of off-farm labor (see Figure 4.2). The Southeast region has the fastest growth rate of labor productivity and off-farm laborers. In 1986, the off-farm labor in the suburbs of Shanghai, Beijing, and Tianjin was more than 50% of the total rural labor force (63.2%, 54.6%, and 48.0% respectively). The coastal provinces (Jiangsu, Zhejiang, Guangdong, and Liaoning) had the second highest proportion of off-farm labor (China's Statistical Yearbooks, 1981 to 1987). In fact, rural industry has become the main income sources in these regions.

Much of the literature on economic development assumes an abundance of labor in agriculture, the marginal product of which is close to zero. W. Arthur Lewis argued that the withdrawal of unproductive labor from agriculture does not cause a significant decline in agriculture. This argument is confirmed by the Chinese agricultural sector (Table 3.1 shows that the marginal product of labor in the agricultural sector is very small after 1980). Although rural industry and other off-farm activities absorbed almost 20% of agricultural labor, total production increased 50% between 1980 and 1985.

Land Productivity Growth

In economic development, labor probably is the most interesting and significant input for traditional agriculture. In land-limited countries, such as China, agricultural production mainly depends on the quantity and quality of land input. The range of possibilities for land utilization and agricultural production therefore are delineated by the major geo-environmental parameters of topography, climate, and soils. Within this range, the actual patterns of land use are determined by a number of factors, such as demand for agricultural products, technology available, and land-labor ratio. Although most regions may be able to increase production through increased labor input (i.e., working harder), the resulting increase in production normally is very low. Use of off-farm inputs, such as fertilizers, pesticides, and other chemicals, and the construction of irrigation systems provide potentials for increased production.

Of China's total area, no more than 11% or about 100

million hectares are arable. Between 1949 and 1978, about 20 million hectares of land were reclaimed for farming (Hsu, 1982) but almost the same amount of arable land has been lost to urban expansion, road and factory construction, and the encroachment of desert. Hence, multiple cropping is widely practiced within these limitations. Early maturation and cultivars of crops with high and stable yields are a continuing priority. Varieties with short growing periods, which are tolerant of either early spring or late autumn cold injury, have been selected. Appropriate varietal combinations and crop rotation patterns have been developed to suit local seasons and conditions. Ingenious methods of interplanting have been devised, such as planting cotton, maize, or watermelons in spaces between rows of wheat or rapeseed and harvesting their crops before the wheat or rapeseed matures. Although the growth of the second crop seedings may lag for a short time, the benefits derived from the mother crop--as a windbreak--may outweigh the temporary competition for water, nutrients, and sunlight (Sun 1987).

Both interplanting and transplanting are labor intensive, and are rational approaches in substituting time for space where land resources are limited and the labor force is immense. Transplanting is an important technology for gaining time and hence increasing cropping index in crop production. This is widely used for rice, cotton, sweet potatoes, and many vegetables. However, the large quantity of fertilizer inputs and improved irrigation facilities are required for the further increase of the cropping index (Sun, 1987).

In Table 2.3, the increases in the cropping index reflect increases in the application of fertilizer and the great efforts made to improve irrigation facilities in all regions; these inputs contributed greatly to agricultural production. (The various sources of land productivity differences among regions and over time are discussed in Chapter 5.)

Land productivity for the different regions is shown in Table 4.2 and Figure 4.3. For the whole country, the growth rate averaged about 5% a year from 1965 to 1986. At the beginning of the period, interregional differences in land productivity were as large as labor productivity. For example, in the South region in 1965 land productivity was 2.03 times the productivity in the Northwest region (70.7 to 34.8 yuan per mu). The same approximate proportions were maintained for 20 years in these two regions. But land productivity has been increased to 207.8 yuan per mu in the South and only to 75.9 yuan per mu in the Northwest. The former was 2.74 greater than the latter. The differences have grown. The position of the Southeast has climbed from the third rank to the second, that of the Southwest has dropped from the second to the fourth.

Most regions doubled their growth rates in land productivity after 1979. In the country as a whole, growth rate was 4% a year before the reform and 7.65% a year afterward. In the South region, the growth rate was stagnant before the reform (2.53%) but very high

Table 4.2 Land Productivity for Different Regions
(Yuan Per Mu of Agricultural Land)

	N.E.	N.	N.W.	C.	S.E.	S.W.	S.	National
1965	40.8	37.6	34.8	56.7	58.9	65.5	70.7	49.5
1970	48.5	47.8	36.4	66.8	70.2	66.5	77.5	57.6
1975	67.6	66.6	45.4	81.6	83.8	73.7	75.6	71.3
1976	63.8	63.6	42.1	77.3	89.8	69.8	88.1	70.1
1977	64.2	63.7	39.0	76.2	83.4	74.3	94.4	70.1
1978	72.9	69.1	41.1	77.4	93.7	82.8	99.5	76.5
1979	73.3	73.0	43.9	92.0	107.4	87.9	96.8	82.5
1980	81.5	80.6	44.6	90.6	106.9	98.3	103.8	87.3
1981	83.8	84.0	51.2	98.9	118.1	102.7	112.4	93.4
1982	89.4	94.7	56.5	110.3	132.0	114.7	129.9	104.6
1983	107.3	106.3	60.8	114.5	136.1	125.2	137.7	113.4
1984	115.5	119.0	68.5	128.2	156.5	136.8	150.3	126.3
1985	106.7	123.1	76.7	137.5	164.6	141.1	170.2	131.9
1986	110.1	116.2	75.9	141.8	181.7	135.4	207.8	136.0
Annual Growth Rate %								
1965 to 79	4.61	5.23	1.81	3.80	4.73	2.29	2.53	4.00
1980 to 86	5.14	6.31	9.28	7.76	9.23	5.47	12.26	7.65
1965 to 86	4.84	5.52	3.78	4.47	5.51	3.52	5.32	4.93

1. Land productivity is measured in terms of yuan per mu, with 1980 constant prices. Land input is measured in terms of sown areas. Grassland is transformed to the sown area equivalence (for details, see Appendix 1).

Figure 4.3 Land Productivity Index

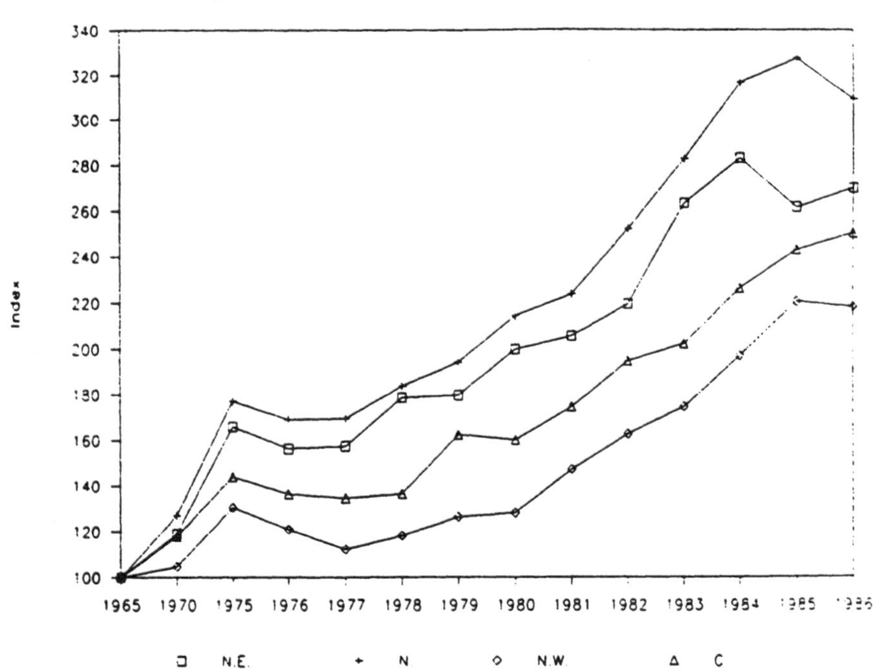

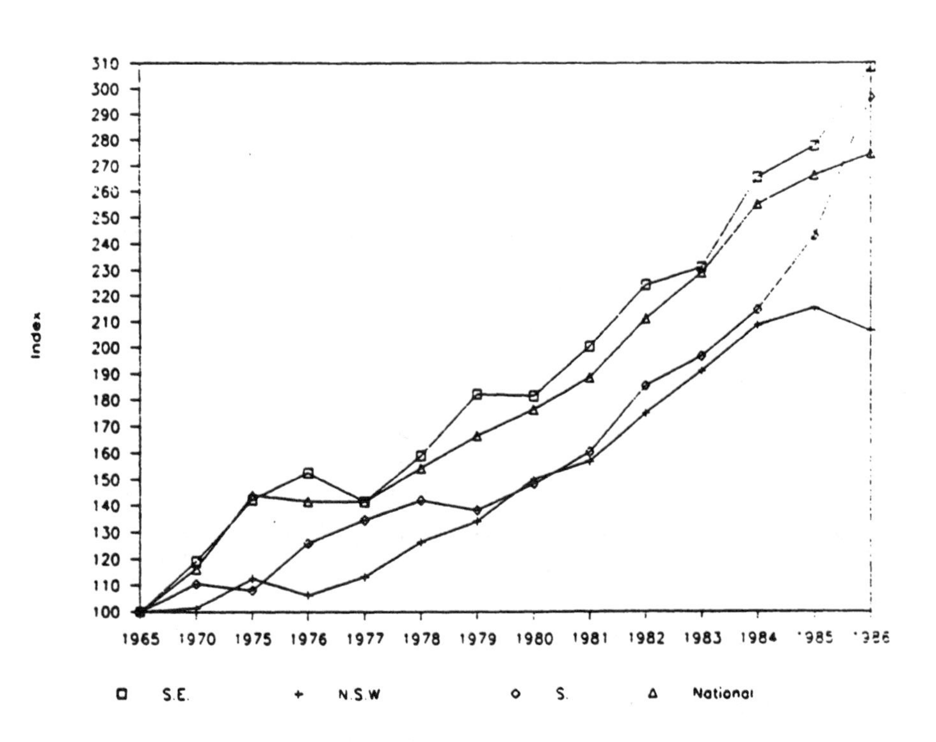

subsequently (12.26%). It was not until 1979 that land productivity in the Northwest began to increase rapidly (1.81% vs. 9.28%). Central, Southeast, and Southwest regions more than doubled their growth rates after 1979. In the North and Northeast, which had high growth before 1979, the growth rates after 1979 increased slightly.

4.3 Total Factor Productivity Growth: Measurement and Comparison

Inasmuch as in Chinese agriculture the rate of modern inputs has far exceeded the growth rate of the sector's total production, should we conclude that no technological progress has been made? According to Wiens (1982), such a conclusion cannot be made.

1. It is difficult to know what the quantitative impact of the law of diminishing returns should be, given the fixed or declining areas of cultivation. Were technological progress absent, perhaps the growth of total production would have been smaller.

2. Technological change is often required merely to permit the increased use of inputs. China could not have extended its irrigated land without changes in the sources of water and in the means of drawing and applying it. Nor could the multiple cropping rate have increased without the development of early maturing varieties which make it possible to squeeze more crops into the frost-free periods.

3. Modern inputs, such as chemical fertilizer and

machinery, supplement or substitute for traditional inputs of land, manurial fertilizer, draft animal, and labor.

Given these reasons, appropriate measures of total factor productivity is necessary for further analyses. In the preceding section it is found that productivity levels and the rate of change in the partial productivities differ by region and time. Some regions have high growth rates in labor productivity whereas other regions have high growth rates in land productivity. In order to compare the overall productivity levels of each region during different periods, a comparison of total factor productivity is needed. Therefore, we discuss first, production elasticities and factor shares to determine the degree of disequilibrium in agricultural production and then, based on the measurement of total productivity, we analyze the differences among regions over time.

Production Elasticities vs. Factor Shares and Disequilibrium

In Chapter 3, the production functions estimated there provide the necessary information for production elasticities. From (3.4), production elasticity for input i can be derived as follows:

$$E_i = \partial \ln(y)/\partial \ln(x_i) = a_i + a_{it}t. \qquad (4.2)$$

We define factor saving and factor using as follows:

if $\partial(E_i/E)/\partial t > 0$, factor i using;

if $\partial(E_i/E)/\partial t = 0$, factor i neutral;

if $\partial(E_i/E)/\partial t < 0$, factor i saving;

where $E = \Sigma E_i$.

Obviously, if $a_{it} > 0$ in (4.2), production elasticity of i is increasing and, consequently, production is i factor using. If $a_{it} < 0$ in (4.2), production elasticity of i is decreasing and, therefore, production is factor saving. If $a_{it} = 0$ in (4.2), production elasticity remains the same and, therefore, production is factor i neutral.

Using functional form (3.7) and the estimates of R3 in Table 3.2 (Chapter 3), the production elasticities for different years can be shown in Table 4.3. The production elasticities of traditional inputs--land, labor, and manurial fertilizer--are decreasing: labor production elasticity at the rate of 3.2% a year; land, 3.8%; and manurial fertilizer, 3.8%. The speed of increase of production elasticities of modern inputs is more rapid than the speed of decrease of traditional inputs. Production elasticity of machinery increased by 14.7% and chemical fertilizer by 2.8% a year. The elasticities of draft animals and irrigation are negative in some years, which, in fact, is not the case. These two elasticities are very close to zero and not significant econometrically.

The State Statistical Bureau in China conducted a survey on the cost shares of each input for different crops. According to their findings, the cost shares of each input for aggregate production are calculated as in Table 4.4.

Although agriculture in China is increasing the use of modern inputs, the shares of traditional inputs are

Table 4.3 Production Elasticities for Different Inputs, 1965 to 1986

	Labor	Land	C.F	Machinery	M.F.	D.Animal	Irri.
1965	.410	.234	.147	.0695	.219	.0554	.0102
1970	.364	.226	.181	.124	.221	.065	.0072
1975	.312	.220	.210	.169	.204	-.016	.0047
1976	.301	.219	.216	.178	.203	-.022	.0042
1977	.290	.217	.221	.187	.202	-.029	.0037
1978	.280	.216	.227	.196	.200	-.035	.0032
1979	.270	.215	.233	.205	.199	-.042	.0042
1980	.260	.214	.239	.213	.198	-.048	.0022
1981	.250	.212	.244	.222	.196	-.050	.0017
1982	.239	.211	.250	.231	.195	-.061	.0012
1983	.228	.210	.256	.241	.194	-.074	.0007
1984	.218	.208	.262	.249	.192	-.080	.0002
1985	.208	.207	.267	.258	.191	-.086	-.0003
1986	.197	.206	.272	.267	.190	-.091	-.0008

1. C.F.: Chemical fertilizer; D. draft: Draft animals. Irri.: Irrigation.

2. Production Elasticities are derived from production function (3.7) rather than from production functions for each year in order to avoid random shock on production elasticities.

Table 4.4 Factor Shares, 1965 to 1986

Year	Labor	Land	Mach.	Ferti.	Irri.
1965	.461	.385	.044	.095	.015
1970	.438	.378	.054	.110	.020
1975	.416	.363	.066	.128	.027
1976	.412	.358	.069	.131	.030
1977	.408	.352	.072	.135	.033
1978	.404	.347	.075	.139	.035
1979	.400	.342	.078	.143	.037
1980	.396	.336	.081	.147	.040
1981	.392	.330	.084	.151	.043
1982	.388	.322	.089	.156	.045
1983	.384	.318	.091	.160	.047
1984	.380	.315	.095	.170	.050
1985	.376	.297	.100	.175	.052
1986	.372	.290	.103	.180	.055

1. Factor shares in 1984 are aggregated from factor shares for different agricultural activities published in Agricultural Yearbook, 1986. Weights are chosen as follows: .35 for rice; .20 for wheat; .20 for cotton and oilseeds; .25 for others.

2. Profit per mu for crops is used as the proxy for land shares.

3. Machinery cost includes both machinery and draft animal cost.

4. The factor shares, except for 1984, are interpolated from the growth trend of cost shares published in Handbook of Economic for Technology (1984).

still large. The sum of land and labor shares is more than 50% for all crops; for labor intensive crops, such as cotton, the share is even greater than 80%. Machinery input share is less than 10% and fertilizer share is about 18%. Irrigation has a dramatically important role in production. The cost of constructing water control systems and irrigation was not included in the total cost of production because in most areas irrigation is regarded as a public good and the central and local government budget pays the huge cost. The government used food subsidies as incentives for surplus labor in winter to construct water control projects and irrigation systems.

The differences between production elasticities and factor shares are very large (comparing Table 4.3 with Table 4.4). For traditional inputs, the factor shares are larger than the production elasticities whereas for modern inputs the factor shares are smaller than their production elasticities. The differences stem from three different sources:

1. Estimation of production function. Mis-specification of production function and inappropriate econometric procedure will bias the estimation of production elasticities.

2. The sample in the survey of factor shares. The data are from 29 different provinces and autonomous regions. There are many qualifications on the accuracies of the data.

3. Disequilibrium in input and output markets. It deserves more detail discussion because it is a common

problem in socialist countries and it has a fundamental effect on the difference between factor shares and production elasticities.

At equilibrium in both input and output markets, the production elasticity for a certain input is equal to its factor share, otherwise, disequilibrium exists. The degree of disequilibrium can be defined as the distance between the factor share and production elasticity.

$$d_i = \frac{e_i - s_i}{s_i} \tag{4.3}$$

where d_i is the relative degree of disequilibrium for input i; s_i is factor share of input i; and e_i is production elasticity of input i. Figure 4.4 presents the disequilibrium for each input in Chinese agriculture from 1965 to 1986.

The disequilibria of labor and machinery show increases over time; fertilizer remained the same; and land input moved close to the equilibrium point. The disequilibrium mainly came from four sources: (1) self-sufficiency strategy, (2) non-existence of the market, (3) irrational campaign in agricultural production, (4) quota system in food procurement.

China's national and regional self-sufficiency policy may cause the inefficient allocation of resources. Each region is expected to produce everything it needs, but the region may not have a comparative advantage in the production of such crops. Self-sufficiency can be achieved either by price or quantity control, but in China the self-sufficiency strategy is achieved mainly by

Figure 4.4 Comparison of Factor Shares and Production Elasticities

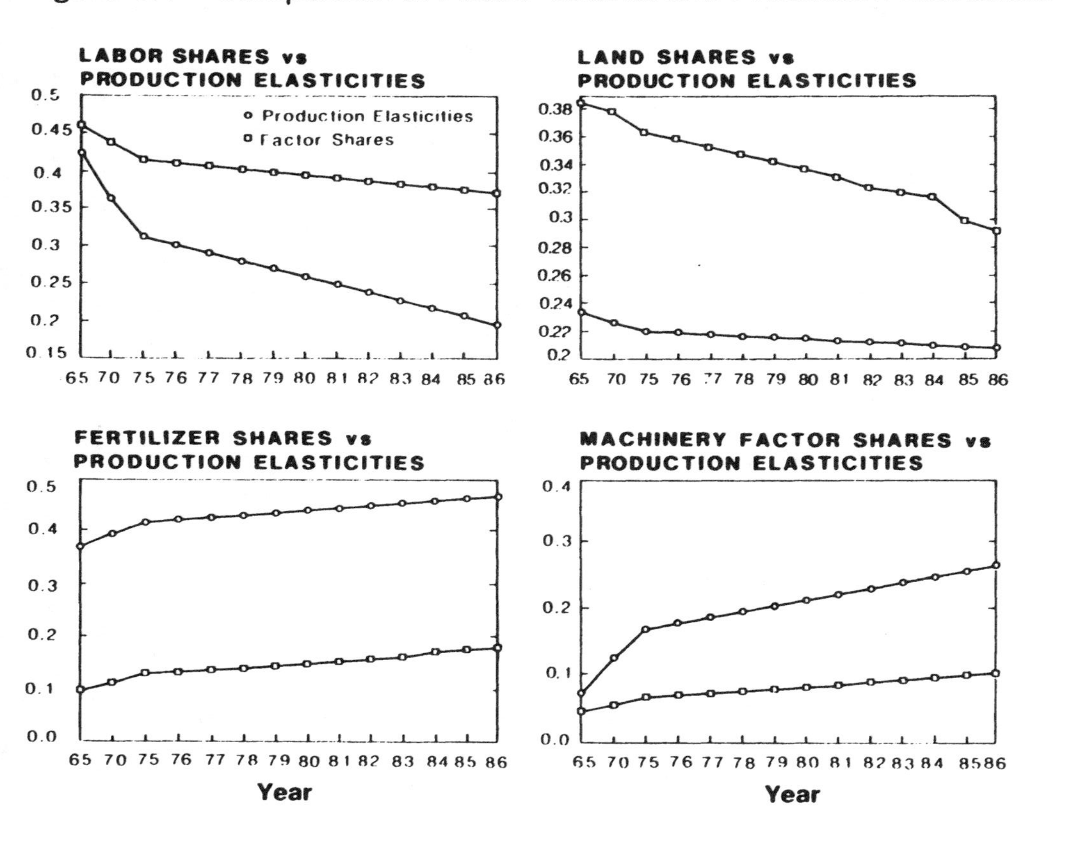

quantity control. As Lardy noted (1983, p.29), the use of quantity rather than price controls has several important implications: (1) The use of quantity controls shifts the burden of achieving self-sufficiency from the state to the peasantry. (2) The use of quantity controls tends to obscure the high cost of attempting to achieve self-sufficiency. (3) Inasmuch as self-sufficiency was imposed at the local rather than national level, the costs of the policy were borne unequally by different regions. Peasant incomes in grain-deficient regions declined relative to incomes of peasants in regions with a comparative advantage in grain production because their pattern of production relative to their pattern of consumption, in which grain has a heavy share, was more specialized.

The absence of a market for labor and the government restrictions on its mobility causes labor to be used inefficiently in agriculture. The marginal production of labor was very small from the production function estimation while the labor cost share covered almost half the total.

During the last four decades, the Chinese government became accustomed to using campaigns to develop agriculture. However, the campaigns were often too "left." Take mechanization as an example: Appropriate machinery must be chosen and appropriate steps of mechanization taken for consistency with agricultural conditions. China, it must be kept in mind, is a land-scarce and labor-abundant country. Mechanization, therefore, must be consistent with this characteristic. If laborers do not have opportunities to leave agriculture, labor-saving mechanization only increases total cost without increasing output. Also, Tam (1985) indicated that unlike other modern inputs (e.g., chemical fertilizer), agricultural machines are major, indivisible fixed assets that are expensive to run and to maintain. In addition, the effective use of such machines often presupposes certain technical and organizational conditions; and the optimal utilization of certain machinery also may require a certain scale and pattern of operation. Indeed, the purchase and operation of a full range of agricultural machines for exclusive use often are beyond the reach of farmers.

Like the self-sufficiency policy, the quota system in product procurement forces production teams to cultivate what the government wants, which may not make full use of a region's comparative advantage.

Methodology of Measurement for Total Factor Productivity Growth

Productivity is the ratio of output to input. The earliest approach to productivity measurement was based on the ratio between aggregate output and a single input which results in a partial productivity, such as land or labor productivity. Different development paths in different countries and regions could be studied by using partial productivities. Hayami and Ruttan in their pioneering study of the induced innovation theory (Hayami and Ruttan,

1971/1985) used partial productivity changes over time in different countries to test the theory.

A more comprehensive approach to the measurement of productivity is based on total factor productivity. An index number approach to the measurement of total factor productivity was initiated by Tinbergen in the 1940s and was used extensively by both Denision and Kendrick in the 1960s (see Kendrick and Vaccara, 1980). An alternative approach to measuring total productivity (Solow, 1957) requires the explicit specification of a production function; Solow demonstrated that the rate of productivity can be identified with the rate of Hicks-neutral technical change, assuming constant returns of scale and competitive markets. A more recent development within the production function and within both theoretical and econometric implications for the measurement of productivity is the cost function approach which is based on the duality theory (Binswanger and Ruttan, 1978). Meanwhile, more general and flexible specifications of function forms, such as the translog, generalized quadratic, generalized-Leontief, and generalized Box-Cox models, have become available. Another recent development is the nonparametric approach; Chavas and Cox (1987) initiated the use of Varian's nonparametric analysis to measure agricultural productivity for U.S. agriculture. Also, some scholars have used the input-output table to analyze productivity growth in some developing economies.

In order to choose the appropriate approach to measure total factor productivity for Chinese agriculture, there are summarized here the approaches that are often used in the literature.

Index Number Approach. The most popular way to measure productivity is the index number approach. It includes the arithmetic, geometric, and Tornqvist-Theil indices. The formula for the arithmetic index is expressed as follows:

$$I_t = (Y_t/Y_0)/(\sum_i S_i(X_{it}/X_{i0}))$$

where Y_t is total output at year t, X_{it} is i^{th} input at year t, S_i is i^{th} input factor share, Y_0 is the total output at base year, X_{i0} is i^{th} input at base year, I_t is arithmetic index for year t.

The arithmetic index implies a homogeneous and linear production function and needs the assumption of competitive equilibrium. Because of its simplicity, it has been used extensively. Its greatest disadvantage is a linear aggregation of inputs, implying an elasticity of substitution among inputs of infinity.

The formula for the geometric index can be expressed as

$$\dot{A}(t)/A(t) = \dot{Y}(t)/Y(t) - \sum_i S_i \times \dot{X}_i(t)/X_i(t)$$

where $\dot{A}(t)/A(t)$ is the production function shift rate.

The geometric index needs assumptions of neutral technical change and a linearly homogeneous production function, and a competitive equilibrium. Initiated by Solow, the geometric index is used to catch all effects of unexplained growth or the residual for technical change measurement. However, aggregation in the Cobb-Douglas form implies an elasticity of substitution between inputs of one.

Both arithmetic and geometric indices are inevitably subject to the well-known "index number problem." The common index formulas are the Laspeyres formula, which uses base-year weights, and Paasche formula, which uses end-year weights. Ruttan has pointed out that the first tends to underestimate productivity growth whereas the second has the effect of biasing upward the measurement of output per unit of input.

The Tornqvist-Theil index can be expressed as:

$$\ln(I_t) = \frac{(.5(S_{jt} + S_{j0}))\ln(Y_{jt}/Y_{j0})}{(.5(S_{it} + S_{i0}))\mathrm{Ln}(X_{it}/X_{i0})}$$

where Y_{jt} and Y_{j0} are j^{th} output at year t and at year 0 respectively; X_{it} and X_{i0} are i^{th} input at year t and year 0 respectively; S_{jt} and S_{j0} are j^{th} output share in total output at year t and year 0 respectively; and S_{it} and S_{i0} are i^{th} factor share at year t and year 0 respectively. The Tornqvist-Theil index is a weighted sum of growth rates, where the weights are output shares in total output produced and input shares in total values of input used.

Diewert described a quantity index as superlative if it is exact for an aggregate function that is capable of providing a second-order approximation to an arbitrary twice-continuously differentiable linearly homogeneous aggregate function (Diewert, 1976).

Production Function Approach. By introducing a time variable to the production (cost or profit) function, the shifting effect can be captured and a productivity index can be constructed. It does not require the aggregation of inputs or the selection of weights. This qualification is especially meaningful in China where there is no systematic data for prices or costs of input. However, the productivity growth in the production function approach is represented by a smooth time trend. Peterson and Hayami (1977) pointed out that this convention fails when technical progress is, in fact, discrete or cannot be approximated by a statistically manageable function of time. Consider the production function as

$$Y_t = F(X_t, t) + \text{error}.$$

A convenient measure of technical change during period t is $\partial \ln F(X_t, t)/\partial t$, the percentage change in output due to an increment of time. Usually, it is assumed that production function form can provide a second-order approximation to an arbitrary twice continuously-differentiable production function.

Nonparametric Approach. Assume that production technology can be expressed as Y = F(X), where Y = Y(y, A) denotes effective output, and X = X(x, B) denotes effective input, and where A and B are output-augmenting and input-augmenting coefficients, respectively. Assuming a neutral technical change and $B_0 = B_t$, the productivity index for measuring productivity growth from time 0 to t can be interpreted as

$$\frac{(A_t - A_0)}{y_0} + 1.$$

Annual estimates of input and output augmentation can be generated by linear programming. This approach to measuring shifts in the production function has advantages and disadvantages compared to previous approaches. The biggest advantage is that it does not require any restrictive assumption for the functional form of the producer's production function; another is it does not make any assumption for the profit-maximizing behavior of producers. The disadvantages are that (1) it does not work if there is technical regress and (2) the required computation is more complex than in the other procedures because it requires the solving of T linear programs.

Comparison of Total Factor Productivity Growth

The review of methodology in the preceding section indicates that the Tornqvist-Theil index approach is the superior method. It avoids the common index number problems and the need for constancy of production elasticities or factor shares among all inputs. The production function approach measures smooth change of total factor productivity but it ignores the productivity change in any specific year. In order to construct a total factor productivity index for each region, the production function approach requires the estimation of production function for each region. Because panel data for each region have long time series and fewer cross-section observations, some econometrical problem will appear. The nonparametric approach is complex in computation. Thus, the Tornqvist-Theil index approach is chosen to estimate total factor productivity in this study. Given that it uses aggregate output (same as one output case), the total factor productivity is measured as follows:

$$\ln(TFPI_t) =$$
$$\ln(Y_t/Y_0) - \sum_i .5\times(S_{it} + S_{i0})\times\ln(X_{it}/X_{i0}) \qquad (4.4)$$

where $TFPI_t$ is total factor productivity index at time t; Y_t and Y_0 are total outputs at time t and 0 respectively; S_{it} and S_{i0} are i^{th} input factor shares or production elasticities at time t and 0 respectively; and X_{it} and X_{i0} are i^{th} inputs at time t and 0 respectively.

The measurement of total factor productivity is very sensitive to the weights used to aggregate total input, hence three different weights are used to construct the total factor productivity index. Because production elasticities of draft animals and irrigation are unrealistic in Table 4.3, we consider first only labor, land, chemical fertilizer, machinery, and manurial fertilizer as inputs. The results are shown in Table 4.5 and Figure 4.5. The total productivity for the whole country by this measurement increased slightly from 1965 to 1986. This measurement of total factor productivity ignores the draft animal input which plays an important role even today. We do not consider it an appropriate measurement, however.

The second measurement uses the sum of draft animals and machinery horse power as power input to construct the total productivity index (see Table 4.6 and Figure 4.6). Unlike the first measurement, the total factor productivity for the whole country had a positive growth rate.

Table 4.5 TFPI Constructed by Production Elasticities Without Draft Animal as an Input

	N.E.	N.	N.W.	C.	S.E.	S.W.	S.	National
1965	100	100	100	100	100	100	100	100
1970	84.9	97.3	91.3	93.4	97.7	80.1	103.3	95.1
1975	94.0	101.7	89.4	96.7	95.1	68.4	85.5	93.4
1976	84.5	91.3	79.7	87.3	92.3	61.0	91.0	86.6
1977	81.3	86.7	70.6	82.6	81.0	61.1	93.5	82.4
1978	92.0	91.3	72.1	81.9	87.1	65.4	96.9	86.9
1979	90.4	93.3	71.1	91.4	95.2	66.1	89.6	89.2
1980	94.9	97.1	70.9	82.6	88.2	69.6	87.8	88.4
1981	94.9	97.8	79.2	86.9	94.7	69.9	89.2	91.3
1982	96.8	101.7	80.0	92.3	98.6	75.1	96.8	95.6
1983	109.4	108.1	80.1	90.5	97.4	76.5	96.6	98.3
1984	111.2	116.4	86.5	99.3	109.2	81.3	102.3	105.8
1985	100.3	117.3	92.2	102.1	112.9	81.1	110.8	107.4
1986	97.8	106.1	82.9	96.4	115.8	71.9	127.7	103.4
Annual Growth Rate %								
1965-79	-.33	-.49	-2.41	-.64	-.35	-2.92	-.78	-.81
1980-86	.51	1.48	2.64	2.62	4.64	.54	6.44	2.65
1965-85	.15	.62	-.86	-.04	.59	-1.23	.64	.27

1. Growth Rates from 1965 to 1985 are calculated using three-year averages.

Figure 4.5 TFPI Constructed by Production Elasticities

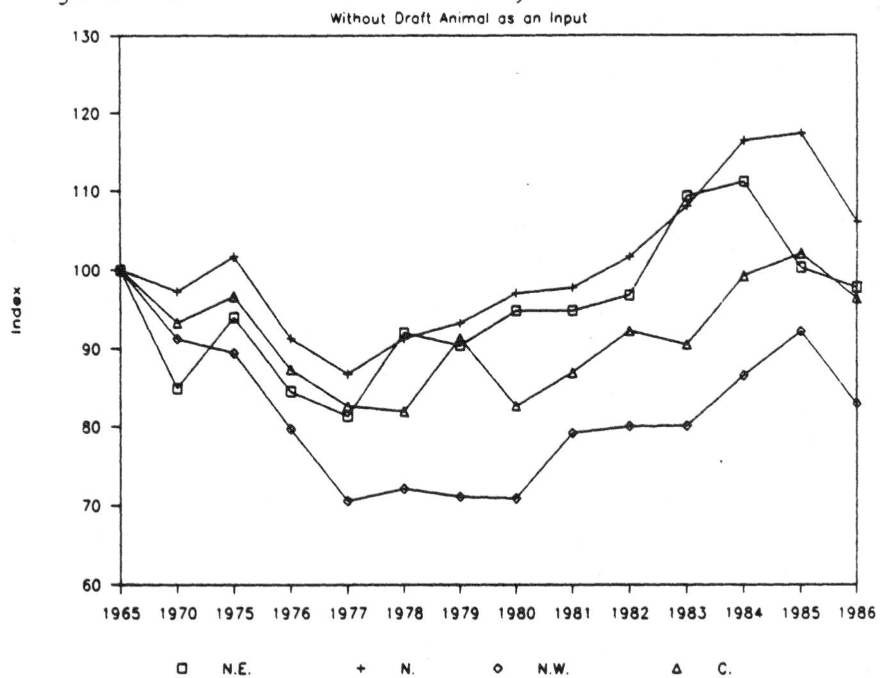

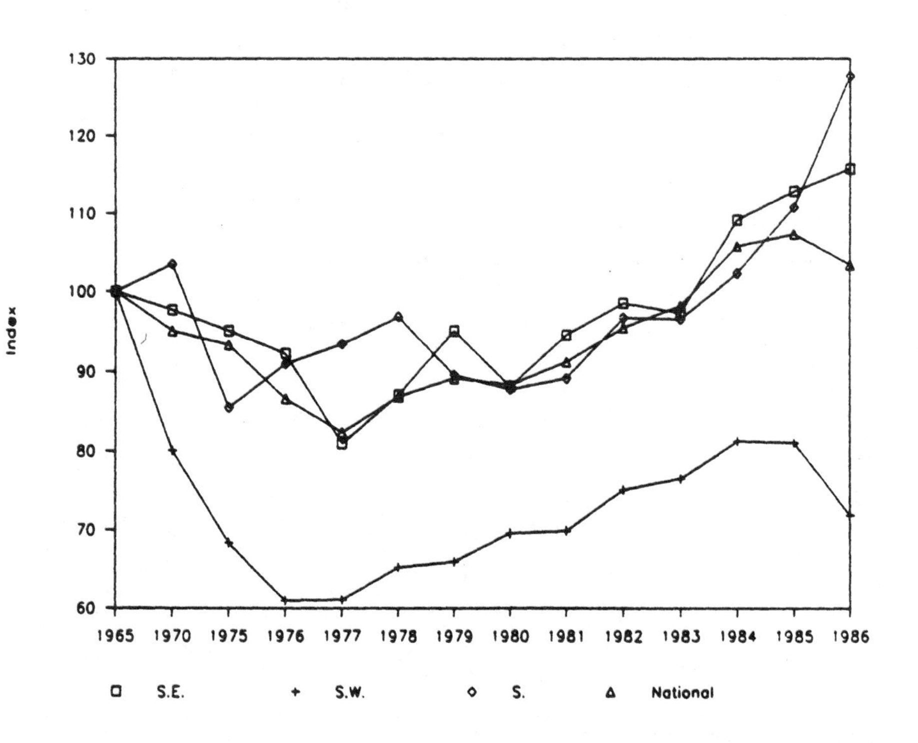

Table 4.6 TFPI Constructed by Production Elasticities With Draft Animal as an Input

	N.E.	N.	N.W.	C.	S.E.	S.W.	S.	National
1965	100	100	100	100	100	100	100	100
1970	88.3	98.6	92.9	97.9	101.8	85.3	107.9	98.2
1975	102.4	111.4	95.2	107.6	106.4	80.2	96.9	103.7
1976	92.6	100.7	85.9	98.2	104.2	72.7	104.3	97.1
1977	90.5	96.4	76.9	93.9	92.1	74.2	108.8	93.3
1978	103.3	102.2	79.2	94.3	99.7	80.8	114.1	99.3
1979	102.1	105.2	79.1	106.2	109.7	83.1	106.4	102.8
1980	108.1	109.9	79.5	96.6	102.1	88.4	105.1	102.4
1981	108.6	111.2	88.6	102.2	110.1	89.4	107.2	106.3
1982	111.2	115.9	89.7	109.1	115.2	96.7	116.9	111.7
1983	126.2	123.9	90.1	107.6	114.4	99.4	117.4	115.6
1984	128.8	134.1	97.5	118.7	128.8	106.5	125.1	125.1
1985	116.8	135.7	104.8	122.8	133.9	107.3	136.1	127.6
1986	114.5	123.2	94.8	116.9	137.9	96.2	158.3	123.6
Annual Growth Rate %								
1965-79	.15	.36	-1.66	.43	.66	-1.32	.44	.19
1980-86	.97	1.92	2.99	3.23	5.16	1.42	7.07	3.18
1965-85	.92	1.36	-.05	.89	1.46	.16	1.69	1.14

1. Growth Rates from 1965 to 1985 are calculated using three-year averages.

Figure 4.6 TFPI Constructed by Production Elasticities

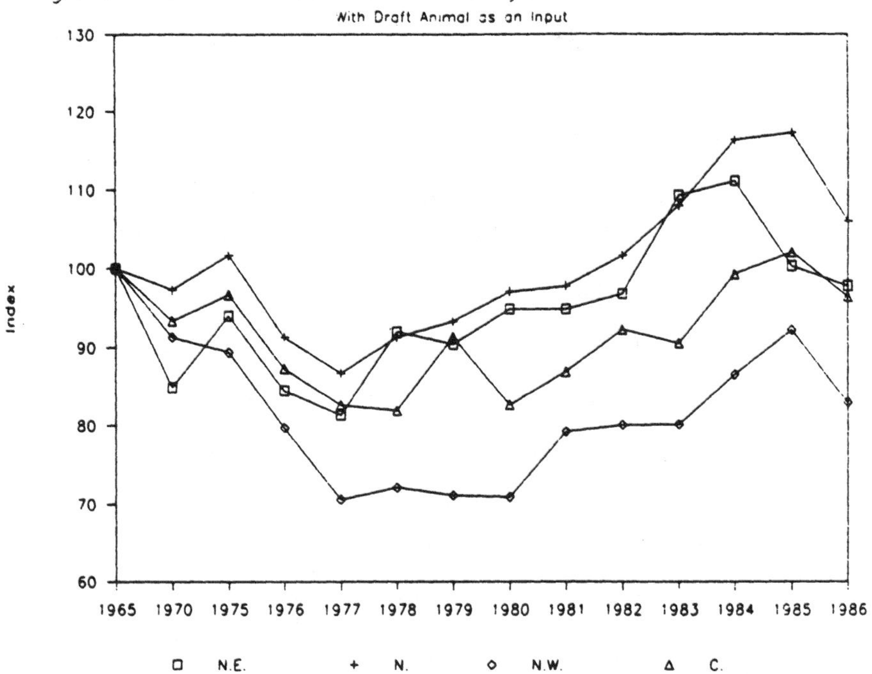

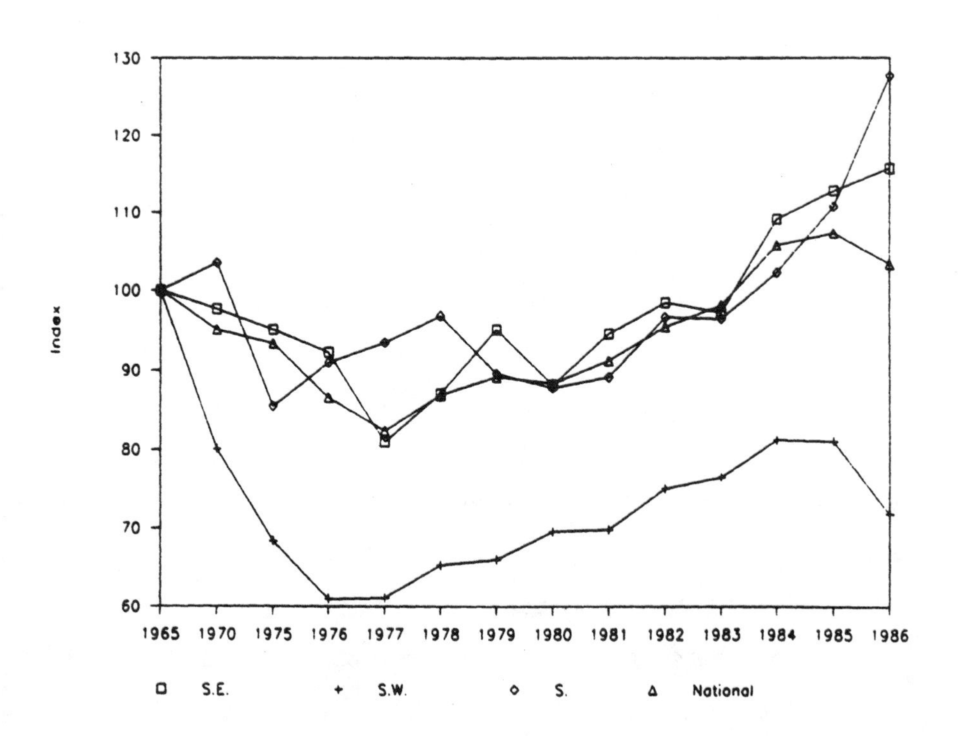

The third total factor productivity index is constructed using the factor shares from the cost survey. The results differ from both previous estimates (see Table 4.7 and Figure 4.7). The total factor productivity for the whole country growth is 3.11% a year with 2.23% before 1979 and 5.15% afterward, and the total productivity increases all the time.

Three different measurements show the great variation using the different weights; after comparing the three, however, the second measurement appears to be more favorable. It does not omit any input information and escapes the disequilibrium problem. Thus, we focus only on the second measurement in the regional analysis.

The total factor productivity was stagnant from 1965 to 1977. Two reasons are responsible: (1) At the beginning of the 1960s industries developed the capability to provide modern inputs and the agriculture began to use more and more of them. At the same time the rural population was growing very fast. Thus the government imposed a restriction on migration of the rural population to other regions or to urban cities. Consequently, the use of traditional inputs did not decrease. (2) The Cultural Revolution, coupled with other erroneous decisions may have caused inefficiency in agricultural production. Whatever technological progress occurred during this period, the changes could not offset the efficiency loss.

After 1980, total factor productivity grew very fast for three reasons: (1) A series of new agricultural

Table 4.7 TFPI Constructed by Factor Shares

	N.E.	N.	N.W.	C.	S.E.	S.W.	S.	National
1965	100	100	100	100	100	100	100	100
1970	101.5	108.1	97.6	105.8	105.9	90.7	106.1	104.1
1975	127.7	138.0	113.7	123.3	121.1	95.1	101.8	120.8
1976	116.9	128.0	105.2	115.3	123.0	88.7	112.4	116.1
1977	117.4	126.0	96.9	113.3	112.4	93.6	119.6	114.9
1978	138.8	139.4	103.8	118.6	127.6	107.2	130.0	127.7
1979	145.7	147.3	107.1	137.4	144.7	113.5	123.5	136.2
1980	157.3	154.8	110.8	127.2	136.1	122.1	123.7	137.6
1981	159.5	158.5	127.8	136.1	151.3	125.2	128.7	145.0
1982	165.1	169.2	133.2	147.8	159.0	136.7	143.2	155.3
1983	190.1	185.5	139.6	148.1	160.4	144.1	146.4	164.0
1984	198.1	204.3	155.4	163.8	182.4	156.1	157.3	180.0
1985	181.7	210.9	171.2	173.5	194.6	159.6	174.6	187.2
1986	180.4	195.4	157.6	171.1	207.4	147.3	207.8	186.0
Annual Growth Rate %								
1965-79	2.72	2.81	.48	2.29	2.67	.91	1.52	2.23
1980-86	2.32	3.96	6.04	5.06	7.27	3.17	9.02	5.15
1965-85	3.17	3.62	2.42	2.67	3.39	2.19	2.98	3.11

1. Growth Rates from 1965 to 1985 are calculated using three-year averages.

Figure 4.7 TFPI Constructed by Factor Shares

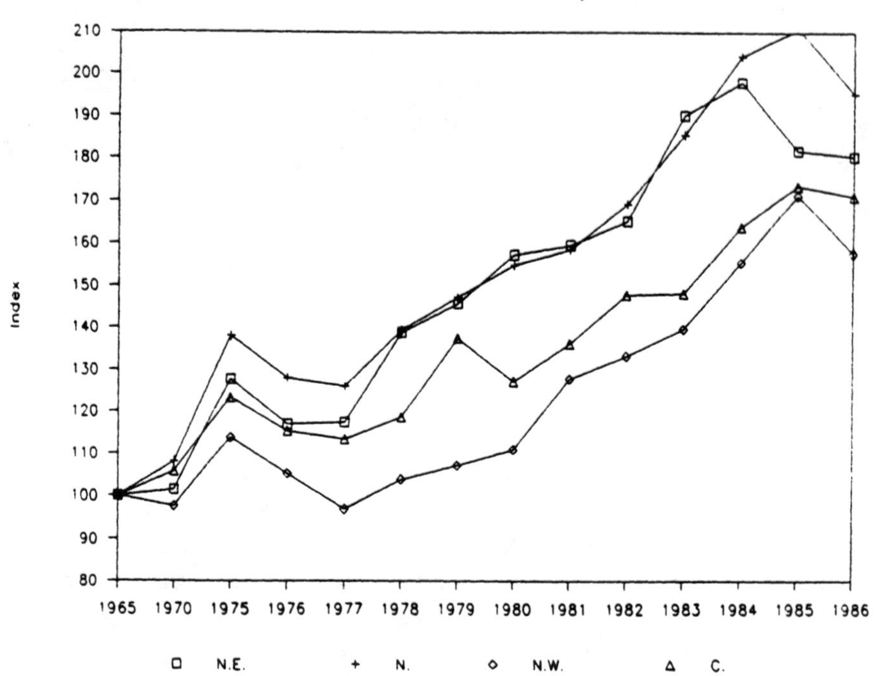

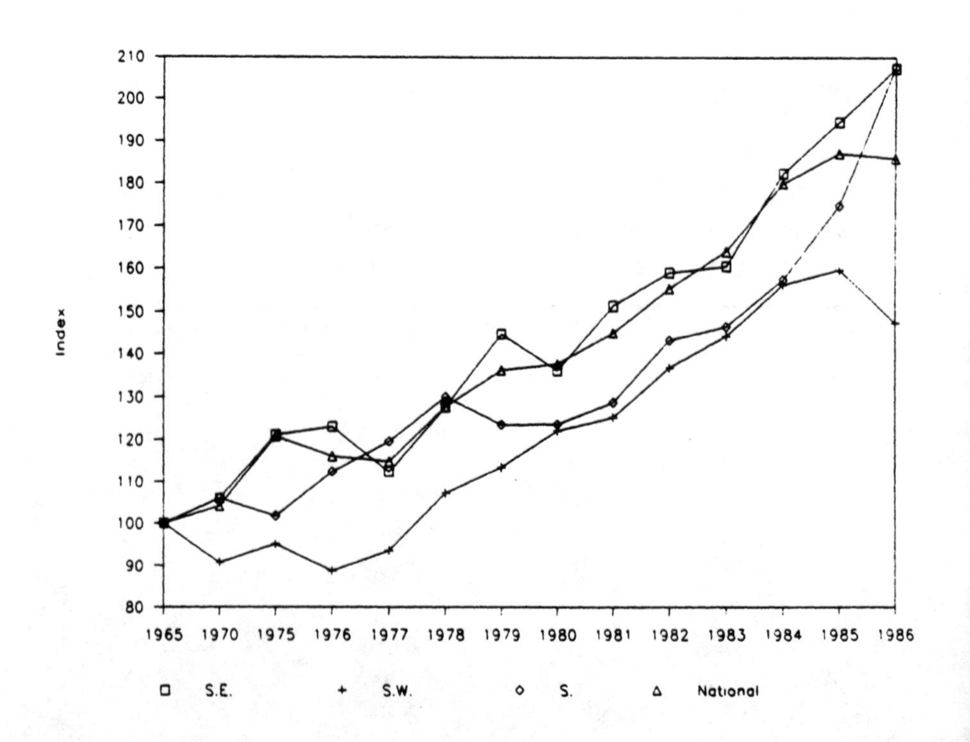

policies resulted in efficiency improvement. (2) Opportunities for off-farm activities became available and lowered the absolute quantity of labor input in agricultural production. (3) The growth rate of modern inputs began to decline while total output continued to increase rapidly.

From 1965 to 1979, despite the successive political movements, most regions had positive growth rates except for the Northwest and Southwest. While, for the whole country as a whole, total productivity increased .19% a year, for the Southwest and Northwest the declines were 1.32% and 1.66% a year, respectively. In other regions, total productivity increased slightly.

From 1980 to 1986, every region showed rapid growth rates. The South and Southeast regions had the fastest growth with 7.07% and 5.16% a year, respectively. In the Central region, the growth was similar to that of the national average while North and Northwest regions had growth rates slightly lower than the national average. The lowest growth rates occurred in the Northeast and Southwest with 1.03% and 1.16% a year, respectively.

During the entire period of 1965 to 1986, total productivity did not increase in the Northwest region. The fastest growth took place in Southeast and South regions with 1.15% and 1.73% a year, respectively. In Northeast, North, and Central regions, the growth rates were similar to that of the national average.

From the provincial analysis (see Appendix 3, productivity growth for each province), the highest growth regions were the suburbs of large cities and the second highest, Guangdong (the earliest market-oriented areas), and Jiangsu provinces; in the latter rural industry has an important role in raising total factor productivity. The lowest growth rate occurred in the provinces and in the autonomous regions with dry and remote areas where water is insufficient for crop production and there is no market for off-farm products. Instead of decreasing the differences in total productivity growth, the reform of the late 1970s increased them.

4.4 Productivity Growth Pattern

In considering the interregional differences in partial productivities of the production factor, which was discussed in the preceding section, it must be noted that the relative positions of individual provinces and regions on the productivity level were not always the same. Some regions preferred land-saving technology, and some did not. It is true that most regions practiced land-saving technology but the speed with which they adopted the technology varied considerably. The preferred combination of specific technology for each region and province must be examined under various conditions, such as factor endowments and other technical and economic conditions, which stem from the historical and geographical backgrounds of the regions.

By making a cross-section comparison of the relations among different partial productivities of agriculture in individual regions, one finds the relative technological preference patterns of each. The comparison also provides insight into the general classifications of several agricultural regions in productivities for comparison with the earlier classifications in Chapter 2 that are based on agricultural and geographical characteristics.

Land Productivity and Labor Productivity

Figures 4.8 and 4.9 show that in 1965 South was the most labor intensive region followed by Southwest, Southeast, Central, and North regions respectively. In relation to the other regions, the Northwest and Northeast are relatively land-using. After 21 years, the situation has changed only slightly. From 1965 to 1986, land productivity grew more rapidly than labor productivity in every region.

The obvious step at this point is to divide the whole period, one part to comprise 1965 to 1979, the other, 1980 to 1986. Figure 4.10 shows the growth rates of labor and land productivities from 1965 to 1979. All regions, except for the Northwest had rapid growth rates of land productivity, which indicates that most regions were practicing land-saving and labor-using technology during those 14 years. The Northeast region alone adopted land-using technology.

After 1980 the situation changed significantly (See Figure 4.11). Because of the reform at the end of 1979 which permits rural labor to engage in off-farm activities, the land-labor ratio has been changed greatly in some regions. Southeast and North regions enjoyed more

Figure 4.8 Labor and Land Productivity Growth Trends for Different Regions

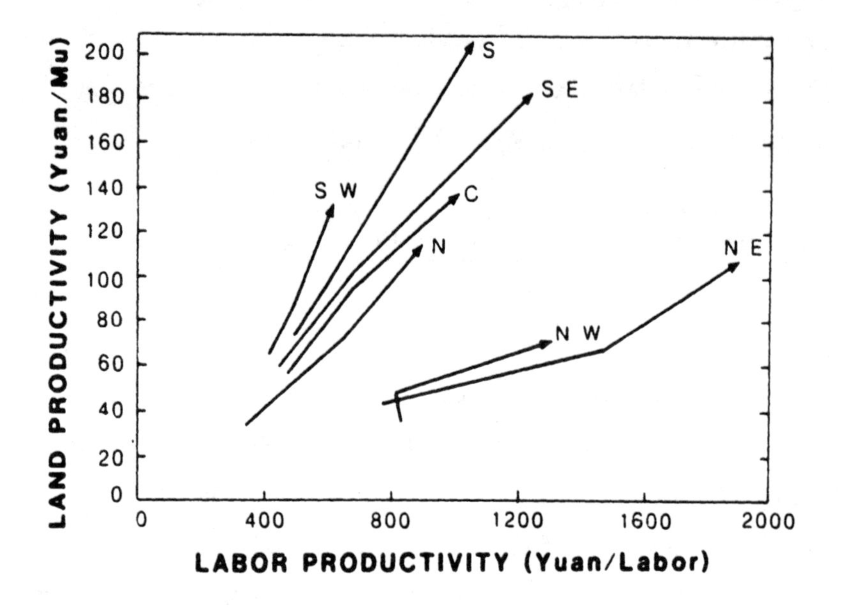

Figure 4.9 Labor Productivity vs. Land Productivity

Annual Growth Rate (%), 1965–86

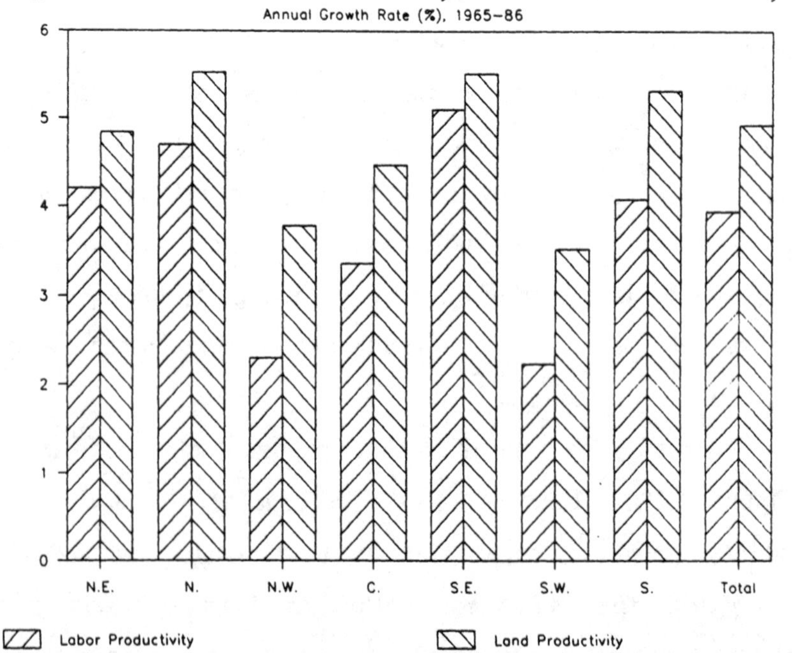

Figure 4.10 Labor Productivity vs. Land Productivity

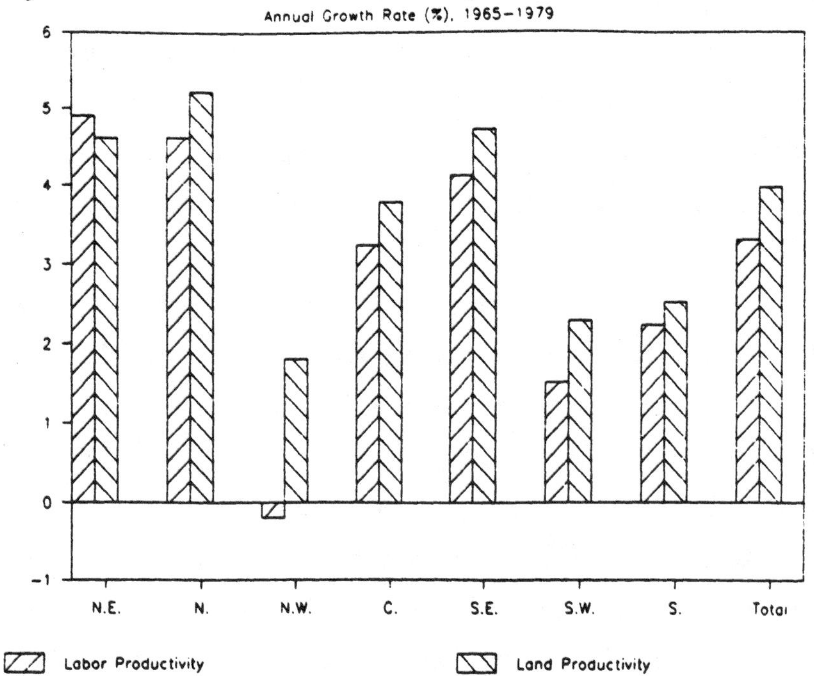

Figure 4.11 Labor Productivity vs. Land Productivity

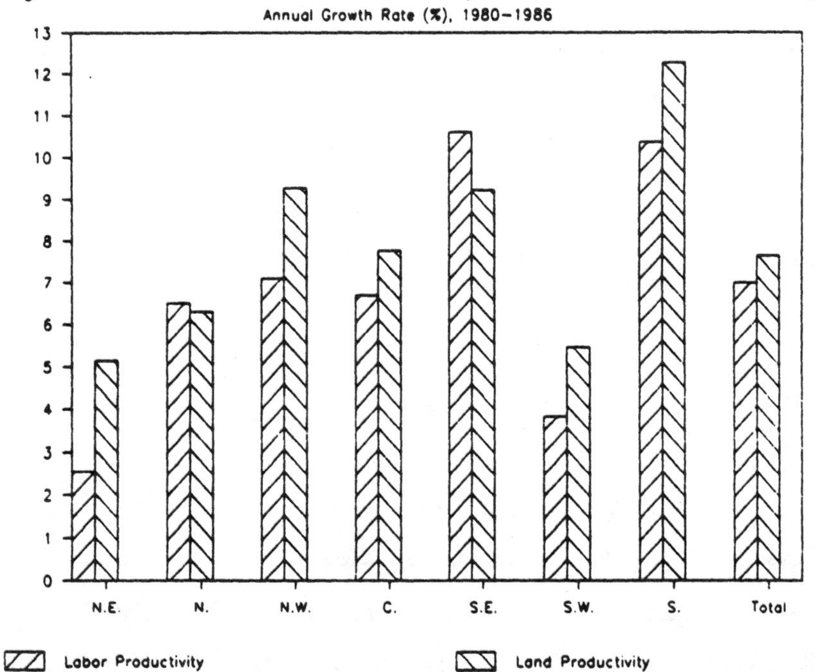

rapid growth in labor productivity than in land productivity. The land productivity growth rate was higher than the labor productivity growth rate in the Northwest region because of the adherence of labor to agriculture. Hence it has changed from a land-using to a labor-using region.

This transition from higher land productivity growth to higher labor productivity growth due to the rural industrial development provides a new development model for developing agriculture (we may call it the rural industrial impact model). Although it is similar to Schultz's (1964) urban-industrial impact model in some sense, the contents are totally different. The rural industrial impact model focuses on rural areas. Labor released from agriculture engages in industrial production in rural areas, and the capital investment for rural industry is also drawn from agriculture surplus. Although the products of rural industry are sold to everyone, including urban consumers or are sold even for export, some products are agricultural inputs, such as chemical fertilizer and farming tools.

Total Productivity and Labor Productivity

The total factor productivity index, we have noted earlier, is in a sense a weighted average of partial productivities of various factors; therefore, the level of partial productivity of each factor for a region must reflect the level of total productivity for that region. However, the relative levels of partial productivity of a factor for individual regions do not necessarily parallel the level of total productivity.

Figure 4.12 shows that the growth of labor productivity is more rapid than that of total factor productivity for all regions. Generally, the higher the rate of labor productivity growth, the higher the total productivity. The correlation between them is not easily identified.

Once again, I divide the 1965-1986 period into two subperiods (1965-1979 and 1980-1986) for analysis. Table 4.13 shows the relation between total productivity and labor productivity before the 1979 reform. For the country as a whole, labor productivity was high during the subperiod but total productivity growth was very slow. The high labor productivity was achieved at the expense of total productivity. Positive labor productivity growth with negative total productivity was found for the Southwest region; implication of these findings is that the high rate of labor productivity growth was achieved by inefficiency in the use of such other inputs as fertilizer or machinery. For the Northeast region, a negative growth rate was found for both labor and total productivity. Here the negative growth rate of labor productivity contributed to the negative growth of total productivity. In the North, Northeast, Southeast, Central, and South regions, the very high labor productivity growth was accompanied by a slight increase in total productivity. The labor productivity did not contribute significantly to total productivity growth.

Figure 4.12 Labor Productivity vs. Total Productivity

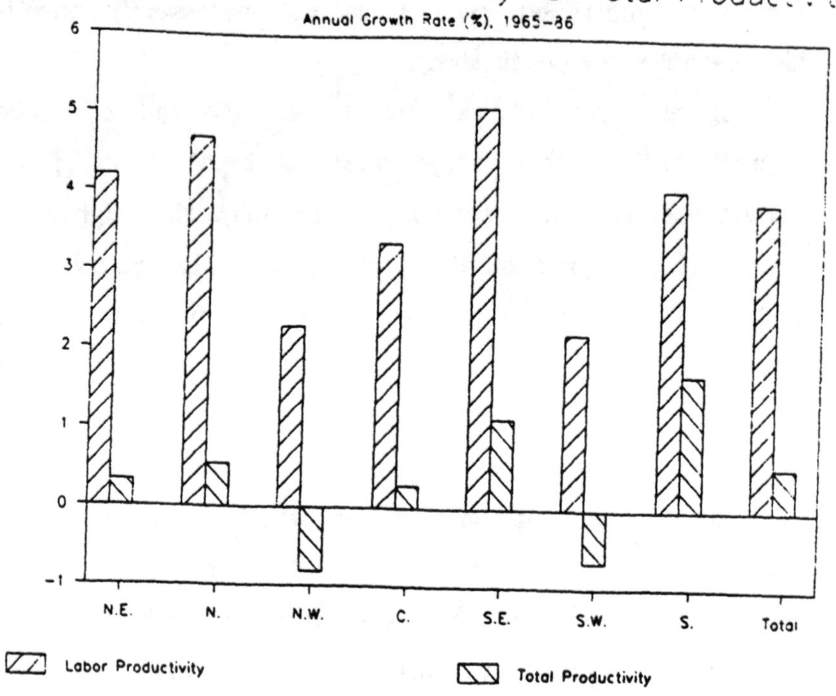

Figure 4.13 Labor Productivity vs. Total Productivity

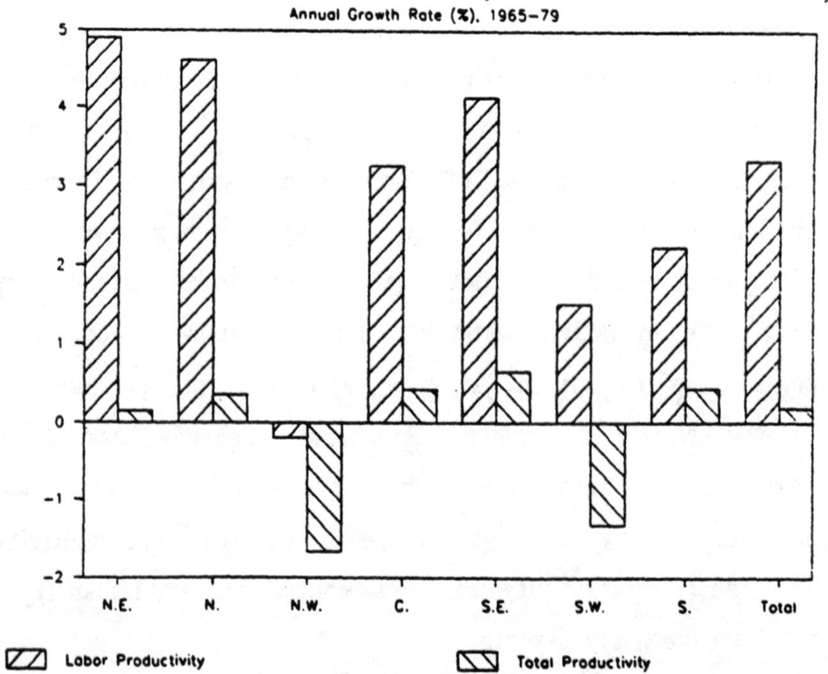

The correlation between total factor productivity and labor productivity after 1980 is stronger than it was before (see Figure 4.14). The decrease of the absolute quantity of labor input raised labor productivity and, therefore, total factor productivity. We can see that the ranks of total factor productivity are the same as those of labor productivity. South region had the highest rank in both total productivity and labor productivity and the Northeast had the lowest.

Figure 4.14 Labor Productivity vs. Total Productivity

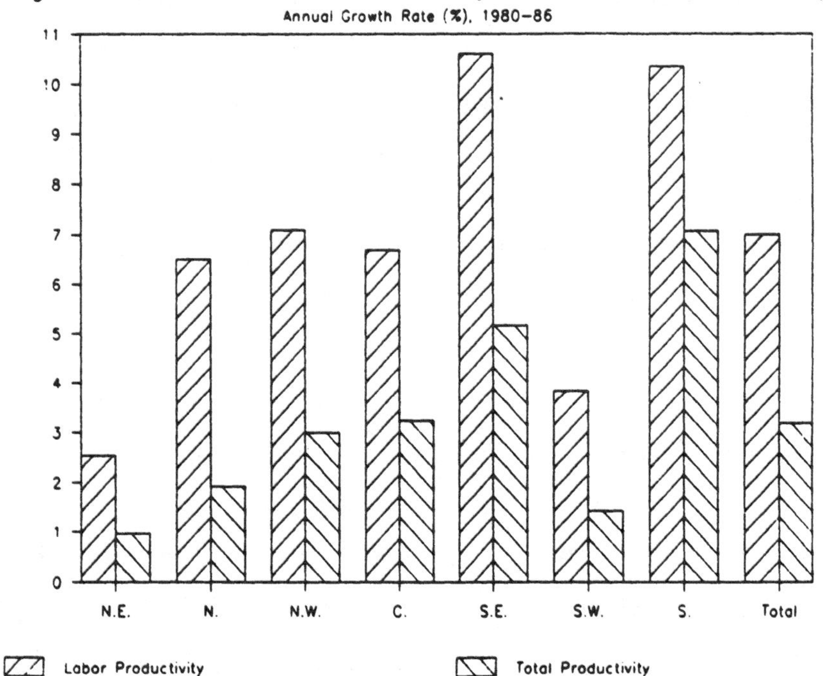

Total Factor Productivity and Land Productivity

Figure 4.15 shows that from 1965 to 1986, the growth rate for land productivity was higher than that for total factor productivity. The difference between the two growth rates is larger than the difference between the growth rates for labor productivity and total productivity. However, in contrast with the relation of total productivity to labor productivity growth, a very high correlation was found between total productivity and land productivity growth (correlation coefficient, .95). This means that the levels of land productivity for different regions clearly reflect the total productivity for the regions. Consequently, the low or negative growth rates of total productivity in the Northwest and Southwest region were caused by the low growth rates of land productivity. In the South, Southeast, and North regions, the higher growth rates of total productivity are related to the higher growth rates of land productivity. These facts imply that in Chinese agriculture technological progress has been biased in favor of land-saving.

Two periods are discussed separately again. From 1965 to 1979, although land productivity grew very fast and total productivity had slow or negative growth, the relation between them was strong (see Figure 4.16). The North and Southeast regions had high land productivity and relatively high total productivity. In the Northwest and Southwest regions, negative total productivity growth was associated with slow growth in land productivity.

Figure 4.15 Land Productivity vs. Total Productivity

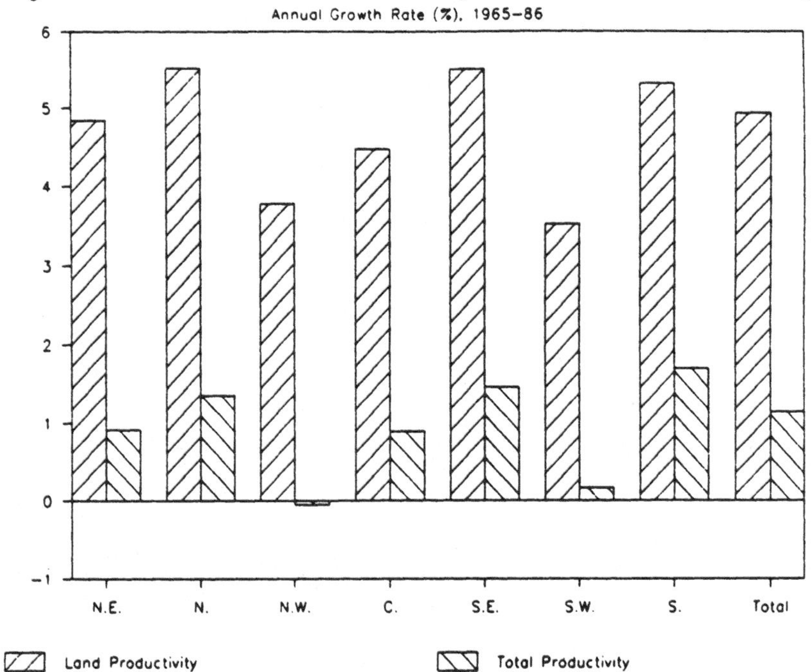

Figure 4.16 Land Productivity vs. Total Productivity

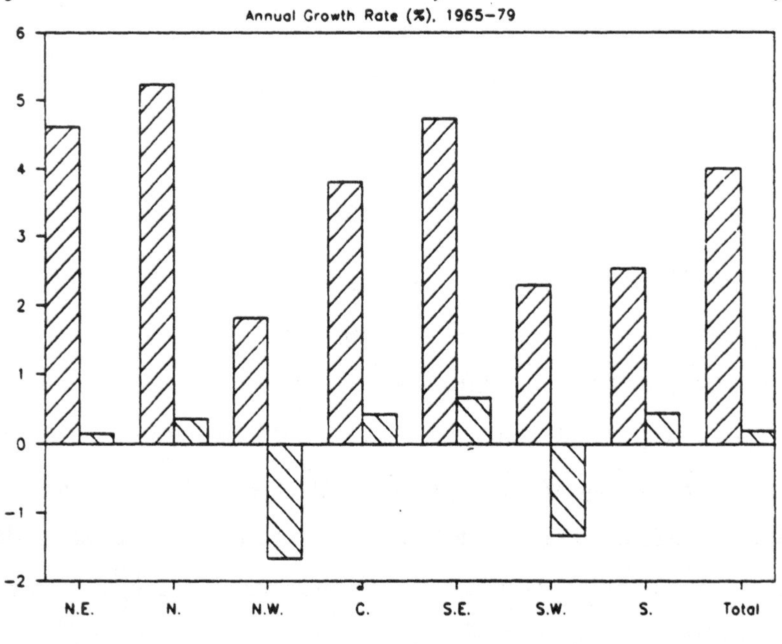

Figure 4.17 Land Productivity vs. Total Productivity

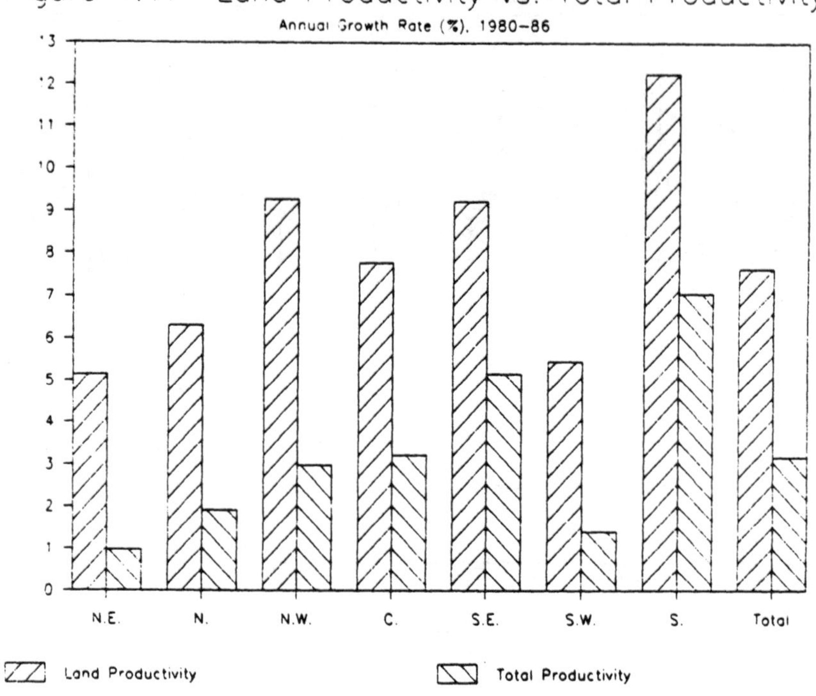

The relation between total productivity and land productivity was even stronger after 1980 than before (see Figure 4.17). The ranks of regions for total productivity were the same as those for land productivity.

4.5 Summary

Based on the measurement and comparison of partial productivities and total factor productivity, the major conclusions in this chapter can be summarized as follows:

1. Both land and labor productivity grew very fast. Nevertheless, land productivity growth was faster than labor productivity growth, which is determined by abundant labor and scarce land resource endowment, and the change in the land/labor ratio.

2. The measurement of total factor productivity is very sensitive to the weights chosen for the aggregation of total inputs. However, every measurement strongly suggests that total productivity for the country as a whole had positive growth rates with its higher rates achieved after 1979.

3. Institutional reform in agriculture after 1979 had a very significant effect on labor, land, and total productivity. On the one hand, the reform reduced the labor inputs in agriculture and, therefore, the total input; on the other hand, the incentives induced by the institutional change made it possible to maintain or accelerate the growth of total production. These two factors contributed to the rapid growth in productivities after 1979.

4. The relation of total factor productivity growth to land productivity growth is stronger than the relation of total factor productivity to labor productivity, indicating that land productivity contributed to total factor productivity significantly and that technical change has been biased in favor of land-saving.

CHAPTER 5
SOURCES OF PRODUCTIVITY GROWTH AND REGIONAL DIFFERENCES

5.1 Introduction: Separation of the Effects of Technical Change and Efficiency Improvement on Productivity Growth

Total production growth results from total input growth and productivity growth. In traditional theory (see Chapter 1), productivity growth is explained as resulting from technical change, assuming that the production unit is technically efficient. However, using the frontier production function, this assumption of a producer's perfect efficiency can be relaxed.

In this chapter, we distinguish between the two sources of total factor productivity change: (1) technical change, which is defined as the shift of the production frontier--maximum output (or potential) the producer could achieve--and (2) efficiency improvement, which is the movement of realized output close to the frontier.

The differences between the two are illustrated easily, as in Figure 5.1. For the sake of simplicity, assume that there is no input change. At time 1 every

Figure 5.1 Separation of Technical Change and Efficiency Improvement

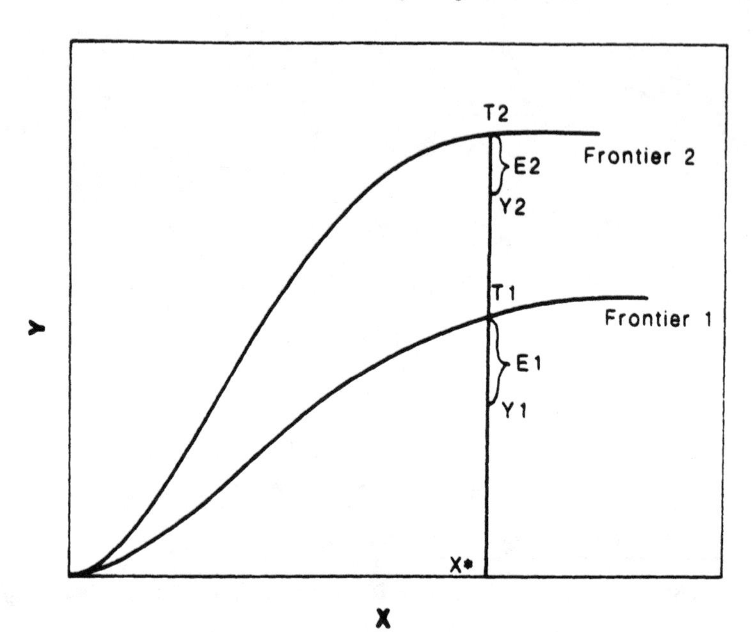

producer faces production frontier 1 and at time 2, frontier 2. However, given the input of x^*, producer's realized output is Y_1 at time 1 and Y_2 at time 2 rather than T_1 and T_2 due to production inefficiency. Technical change is measured by the distance between the frontiers (i.e., $T_2 - T_1$ using input x^*) and production inefficiency is represented by the distances between the frontiers and realized outputs (i.e., E_1 and E_2). The improvement of efficiency over the time periods, then, is the difference between E1 and E2 ($E_1 - E_2$) and the change in total factor productivity is the sum of the change in production frontier and the change in efficiency:

$$\frac{Y_1 - Y_2}{X^*} = \frac{(T_2 - T_1)}{X^*} + \frac{(E_1 - E_2)}{X^*} .$$

In Chapter 4, we discussed the interrelations among

productivities; in this chapter we examine the sources of productivity growth and regional differences by using the estimation of aggregate agricultural production functions in Chapter 3 as our base. The sum of the estimated coefficients (see Table 3.2) indicates that the assumption of a linear homogeneous production function in Chinese agriculture is not too unreasonable, therefore, the rates of growth and the differences among regions can be expressed as the sum of growth rates or differences in inputs weighted by the relevant production elasticities.

Using function form (3.7), production function can be expressed as:

$$\begin{aligned} \mathrm{Ln}Y(t) &= \ln a_0 + \sum_i a_i \ln x_i(t) + \sum_i a_{it}(\ln x_i(t))\times t \\ &\quad + a_t t + a_{tt} t^2 + \ln(e^{u(t)}) + v(t) \qquad (5.1) \\ &= \ln A_0(t) + \sum_i a_i(t)\ln x_i(t) + \ln E(t) \qquad (5.2) \end{aligned}$$

where $\ln A_0(t) = \ln a_0 + a_t t + a_{tt} t^2 + v(t)$; $a_i(t) = a_i + a_{it} t$; and $\ln E(t) = \ln(e^{u(t)})$. By taking the first derivative of (5.2) with time t, we have the following expression:

$$\begin{aligned} \partial \ln Y(t)/\partial t &= \underbrace{\partial \ln A_0(t)/\partial t}_{(1)} + \underbrace{\sum_i a_i(t)\times\partial \ln x_i(t)/\partial t}_{(2)} \\ &\quad + \underbrace{\sum_i \ln x_i(t)\times\partial a_i(t)/\partial t}_{(3)} + \underbrace{\partial \ln E(t)/\partial t}_{(4)}. \qquad (5.3) \end{aligned}$$

The first term in (5.3) measures neutral technical change. The input change on production growth is captured by term 2. The third term is the substitution effects on production growth. If the term is positive, then output has been increased through biased technical change (substituting abundant resources for scarce resources). The last term reflects the efficiency improvement in production growth.

Similarly, partial productivity change can be accounted for by technical change, increase of other inputs, and efficiency change.

$$\partial(\ln(Y(t)/x_j(t))/\partial t =$$

$$\underset{(1)}{\partial \ln A_0(t)/\partial t} + \underset{(2)}{\sum_i a_i(t) \times \partial \ln(x_i(t)/x_j(t))/\partial t}$$

$$+ \underset{(3)}{\sum_i \ln(x_i(t)/x_j(t)) \times \partial a_i(t)/\partial t} + \underset{(4)}{\partial \ln E(t)/\partial t}. \qquad (5.4)$$

In accounting for partial and total productivity change, neutral and biased technical change are treated as residuals, that is first and third terms in (5.3) and (5.4) are treated as residuals.

5.2 Accounting for Labor Productivity Growth and Differences among Regions

Growth Accounting of Labor Productivity

The results of accounting for the growth in labor productivity between 1965 and 1986 are presented in Table 5.1. The first row shows the percentage change in labor productivity as of 1986. For all China, labor productivity change slowed by 8.8% (percentage of total productivity change i.e., -4.9/55.6) because of the decrease of land input per unit of labor from 1965 to 1986. Rapid population growth--hence the deterioration of

Table 5.1 Accounting for Agricultural Labor Productivity Growth, 1965 to 1986

	N.E.	N.	N.W.	C.	S.E.	S.W.	S.	National
Labor Productivity								
Change(%)	58.0	62.0	38.5	50.0	64.9	37.1	56.8	55.6
	(100)	(100)	(100)	(100)	(100)	(100)	(100)	(100)
Land	-2.9	-3.8	-7.5	-5.5	-1.8	-6.6	-6.2	-4.9
	(-5.0)	(-6.1)	(-19.4)	(-11.0)	(-2.8)	(-17.8)	(-10.9)	(-8.8)
C.Fertilizer	19.5	19.2	18.8	17.8	18.2	18.4	14.4	18.1
	(33.6)	(31.0)	(48.8)	(35.6)	(28.0)	(49.6)	(25.4)	(32.6)
M.Fertilizer	.96	4.8	3.7	1.5	1.7	1.7	1.8	3.0
	(1.7)	(7.7)	(9.6)	(3.0)	(2.6)	(4.6)	(3.2)	(5.4)
Power	13.6	14.3	10.7	13.2	15.2	11.0	12.7	13.6
	(23.4)	(23.1)	(27.9)	(26.4)	(23.4)	(29.6)	(23.4)	(24.5)
Efficiency	-3.3	9.6	6.1	9.6	10.9	1.3	22.4	8.9
	(-5.7)	(15.5)	(15.8)	(19.2)	(16.7)	(3.6)	(39.4)	(16.0)
Technology	30.1	17.9	6.7	13.4	20.7	11.3	11.7	16.2
	(51.9)	(28.9)	(17.4)	(26.8)	(31.9)	(30.5)	(20.6)	(30.4)

1. The parenthetical figures are percentages; the growth of labor productivity is given as 100.

2. C.Fertilizer denotes chemical fertilizer; and M.Fertilizer denotes manurial fertilizer.

3. Machinery includes draft animals. The transfer of draft animals to machinery is shown in Appendix 1.

4. Formula (5.4) is used for calculation.

the land/labor ratio--offset 19.4% of the labor productivity growth in the Northwest. In Southeast region by 1986 only 2.8% of the negative effects of labor productivity were affected by land decrease. The reason for this small effect is that the well-developed rural industry absorbed a large quantity of surplus labor and, consequently, the land/labor ratio has changed significantly since the end of the 1970s. The reclamation of land in Northeast region slowed the decline of the land/labor ratio and therefore the land/labor ratio change only decreased labor productivity by 5%.

Increased use of chemical fertilizer per unit of labor contributed greatly to labor productivity growth. For the entire country, about 32.6% of the labor productivity increase was contributed by increased chemical fertilizer application. The differences of the contribution of chemical fertilizer input per unit of labor to the labor productivity increase varied from 25.4% in the South region to 49.6% in the Southwest. The contribution of manurial fertilizer to labor productivity was not significant for most regions, ranging from 1.7% in the Northeast to 9.6% in Northwest.

Another important factor in labor productivity growth was the increase of power input (including machinery and draft animals). Almost one quarter of labor productivity growth came from the increase of power input per unit of labor. However, in every region, the contribution of power input per unit of labor to labor productivity (ranging from 23.1% in the North to 29.6% in the Southwest) was smaller than that of chemical fertilizer.

In all regions, the increase of efficiency and technical change contributed to labor productivity growth significantly. Almost half the labor productivity growth is accounted for by technical change and efficiency improvement in South, Southeast, Central, North, and Northeast regions; and in the Southwest and Northwest, about one-third of the labor productivity increase was contributed by efficiency improvement and technical change. Technical change had a greater effect on labor productivity than did efficiency improvement; the latter increased labor productivity by 16% whereas technical change increased labor productivity by 30.4%.

Although the differences in labor productivity are large, the sources of productivity growth seem to be similar in all regions. Technical change and efficiency improvement explain half the growth, and increased chemical fertilizer and power use accounts for the rest. However, rigorous analyses of sources in labor productivity growth among regions are needed to make more secure judgment.

Accounting for Labor Productivity Differences among Regions

From the growth accounting, we can only observe sources of productivity change. In this section, therefore, the sources of labor productivity differences among the regions are analyzed. Analyses for two years, 1965 (Table

5.2) and 1986 (Table 5.3), show changes in the sources of productivity differences among the regions over time.

Differences in land endowment per unit of labor play an important role in explaining labor productivity differences among regions in 1965. The high labor productivity in the Northwest and Northeast regions was associated with their favorable land/labor ratios compared to other regions. The low labor productivity in the Southwest and Southeast regions was due mainly to their scarcity of land.

Although increased power input per unit of labor contributed greatly to the labor productivity growth between 1965 and 1986 (See Table 5.1), the differences in power input per unit of labor are not important in explaining the differences in labor productivity in 1965.

At that time, manurial fertilizer was the main source of nutrients for crop planting, thus in Northwest and Northeast regions where livestock numbers are large--and the quantity of manurial fertilizer is consequently large--high labor productivity also was associated with the quantity of manurial fertilizer per unit of labor. Chemical fertilizer input was not important in 1965 so the contribution of differences in chemical fertilizer application was very small compared to the differences in labor productivity, except in the Southwest region. Chemical fertilizer input there was 12.5% lower than the national average which, with the low land input per unit of labor (11.5% lower than the national average), resulted

Table 5.2 Accounting for Agricultural Labor Productivity Differences, 1965 (Regions vs. National Average)

	N.E.	N.	N.W.	C.	S.E.	S.W.	S.
Difference in Labor Productivity(%)	44.0	-29.0	46.6	12.7	-.7	-12.7	2.2
	(100)	(100)	(100)	(100)	(100)	(100)	(100)
Land	12.6	.5	14.6	0	-4.7	-11.5	-8.9
	(28.6)	(-1.7)	(31.3)	0	(671)	(90.6)	(-404)
C.Fertilizer	-.2	-.8	-.8	1.4	2.7	-12.5	.8
	(-.5)	(2.8)	(-1.7)	(11.0)	(386)	(98.4)	(36.4)
M. Fertilizer	6.9	-7.3	17.5	-3.6	-9.4	1.6	-3.3
	(15.7)	(25.2)	(37.6)	(-28.3)	(1343)	(-12.6)	(-150)
Power	3.6	.2	4.9	-.8	-7.3	-1.6	-.4
	(8.2)	(-.7)	(10.5)	(-6.3)	(104)	(12.6)	(-18.2)
Efficiency	21.1	-21.6	10.4	15.7	18.0	11.3	14.0
	(48.2)	(74.5)	(22.3)	(123.6)	(-2571)	(-88.9)	(636)

1. A similar formula is used for both difference accounting and growth accounting:

$$\frac{\Delta \frac{Y}{L}}{\left(\frac{Y}{L}\right)_0} = \sum_i a_i \frac{\Delta \frac{x_i}{L}}{\left(\frac{x_i}{L}\right)_0} + \frac{\Delta E}{E_0}$$

where Δ denotes the difference between comparing region and national average. Y denotes output, L is labor input, x_i is i^{th} input except labor, and E denotes relative efficiency.

2. Negative numbers in parentheses signify that differences in input use or efficiency improvement have a negative effect on labor productivity.

in 12.7% lower labor productivity than the national average.

The differences in efficiency affecting labor productivity among regions were very large. The lowest efficiency in labor productivity is found in the North region where the 75% of productivity difference from the national average is explained by relative inefficiency. North region is the largest in the country in terms of total labor input, therefore the efficiency of the region plays a crucial role in national average efficiency.

Labor productivity differences among regions grew from 1965 to 1986 (see Tables 5.2 and 5.3). In 1965 the lowest labor productivity (North region) was 29% below the national average, and the highest labor productivity (in the Northwest region) was 46.6% above the national average. In 1986, on the other hand, labor productivity in the Southwest region was 59.5% lower than the national average and in the Northeast, 47.1% higher than the national average. Land endowment per unit of labor still plays an important role in explaining the differences in labor productivity in 1986.

In 1986 in the Northeast and Northwest regions, in which labor productivity was 25% and 46%, respectively, higher than national average, the differences are accounted for by land endowment per unit of land. About 20% of the labor productivity difference below the national average is explained by unfavorable land/labor ratio in the Southwest. However, in the South region, with 9.3% lower land per labor than the national average,

Table 5.3 Accounting for Agricultural Labor Productivity Differences, 1986 (Regions vs. National Average)

	N.E.	N.	N.W.	C.	S.E.	S.W.	S.
Difference in Labor							
Productivity(%)	47.1	-10.2	26.1	1.8	20.3	-59.5	5.0
	(100)	(100)	(100)	(100)	(100)	(100)	(100)
Land	11.9	1.2	12.1	-.5	-1.3	-12.1	-9.3
	(25.5)	(-11.8)	(46.4)	(27.7)	(-6.4)	(20.3)	(-186.0)
C. Fertilizer	13.4	.9	-4.8	-.5	5.5	-18.5	-1.3
	(28.5)	(-8.8)	(-18.4)	(-27.7)	(27.1)	(31.1)	(-26.0)
M. Fertilizer	4.6	-3.7	15.3	-5.0	-10.2	0	-4.3
	(9.8)	(36.3)	(58.6)	(277.8)	(50.2)	(0)	(86.0)
Power	14.2	6.6	12.1	-6.6	-1.0	-31.5	-8.5
	(30.1)	(-64.7)	(46.4)	(-366)	(-4.9)	(52.9)	(-170.0)
Efficiency	3.0	-15.2	-8.6	14.4	27.3	2.6	28.4
	(6.4)	(149.0)	(33.0)	(800.0)	(134.5)	(4.4)	(568.0)

See Note 1 of Table 5.2 for the accounting formula.

the labor productivity is 5% higher than the national average.

Power input per unit of labor has become more important in explaining the differences in labor productivity in 1986 in the Northeast, North, Northwest, and Southwest regions. The higher labor productivity in the Northeast and Northwest also was associated with higher power input, whereas the lower labor productivity in the Southwest related to its lower power input. In Central, South, and Southeast regions, given the low power input per unit of labor, labor productivity is higher than the national average.

The differences in chemical fertilizer input per unit of labor had the same effects as power input on the labor productivity differences.

Although the differences have decreased, the efficiency still explains huge differences in labor productivity. Labor production efficiency was higher in Central, Southeast, and South regions and lower in North and Northwest.

By comparing Tables 5.2 and 5.3, one shows that although land input per unit of labor has decreased its role in explaining the labor productivity differences among the regions between 1965 and 1986, it still is a very important factor in affecting labor productivity in different regions. Power input per unit of land did not affect labor productivity very much in 1965, but in 1986 it became one of the main factors. Chemical fertilizer input also increased in importance in accounting for labor productivity differences whereas the role of manurial fertilizer was unchanged. Differences in the efficiency of labor productivity declined among regions between 1965 and 1986. The Southeast changed from a region with lower labor productivity in 1965 to a region in which labor productivity was higher than the national average in 1986; at the same time the efficiency relative to the national average declined. South region has the highest relative efficiency whereas Northwest has had the lowest relative efficiency in both 1965 and 1986.

5.3 Accounting for Land Productivity Growth and Differences among Regions

Land productivity is of primary importance in Chinese agriculture. Because it is limited, the only way to increase agricultural production is to increase land productivity. In this section, an analysis is presented of the sources of land productivity growth and the differences among the different regions.

Accounting for Land Productivity Growth

The sources of land productivity growth between 1965 and 1986 are presented in Table 5.4. Population increased rapidly in every region over time as did labor input per unit of land. Given the rather large surplus of labor in rural areas, the increased labor input did not contribute to land productivity greatly due to the diminishing return of labor input. The growth of both power and chemical fertilizer input per unit of land contributed to land productivity growth significantly; however, the contribution of chemical fertilizer per unit of land was the greatest. Manurial fertilizer has had almost the same effect on land productivity growth as has labor input.

Technical change and efficiency improvement explains 30% of the change in land productivity. Unlike labor productivity, the effects of efficiency improvement are greater than those of technical change. Among all regions, Northeast is the only region where efficiency has deteriorated while technical change contributed to land productivity greatly. The South had the highest efficiency improvement, however there was a technical regress in land productivity.

Accounting for Land Productivity Differences among Regions

The sources of differences in land productivity between the various regions and the national average for 1965 are shown in Table 5.5. In 1965, labor was an important input in agricultural production. The low land productivity in the Northwest and Northeast mainly came from low labor input while in the South, Southwest, and Southeast high labor productivity was achieved by labor intensive farming. Unlike the difference in labor productivity, the differences in fertilizer input per unit of land contributed to differences in land productivity greatly. In the South, Southeast and Southwest regions, the higher land productivity was associated with more chemical fertilizer input per unit of land whereas in

Table 5.4 Accounting for Land Productivity Growth from 1965 to 1986

	N.E.	N.	N.W.	C.	S.E.	S.W.	S.	National
Land Productivity								
Change(%)	63.0	67.7	54.2	60.0	67.6	51.6	66.3	63.6
	(100)	(100)	(100)	(100)	(100)	(100)	(100)	(100)
Labor	3.7	4.6	7.9	6.3	2.4	7.2	6.9	5.7
	(5.9)	(6.8)	(14.6)	(10.5)	(3.5)	(14.0)	(10.4)	(9.0)
C. Fertilizer	19.7	19.4	19.3	18.4	18.4	19.0	15.8	18.6
	(31.3)	(28.7)	(35.6)	(30.7)	(27.2)	(36.8)	(23.8)	(29.2)
M. Fertilizer	3.3	7.1	8.0	5.3	3.1	6.0	5.9	6.1
	(5.2)	(10.5)	(14.7)	(8.8)	(4.6)	(11.6)	(8.9)	(9.6)
Power	14.0	14.7	12.3	13.9	15.3	12.4	13.6	14.2
	(22.2)	(21.7)	(22.7)	(23.2)	(22.6)	(24.0)	(20.5)	(22.3)
Efficiency	-3.6	10.5	8.6	11.5	11.3	1.9	26.1	10.2
	(-5.7)	(15.5)	(15.8)	(19.2)	(16.7)	(3.6)	(39.4)	(16.0)
Technology	25.9	11.4	-1.9	4.6	17.1	5.1	-2.0	8.8
	(41.1)	(16.8)	(-3.5)	(7.7)	(25.3)	(9.8)	(-3.0)	(13.8)

1. Accounting formula is the same as for labor productivity.

North and Northwest, the low productivity was related to less chemical fertilizer application. Manurial fertilizer input also had a positive correlation with land productivity in most regions. Differences in power input per unit of land had little effect on the differences in land productivity.

The differences in efficiency affecting land productivity were large in 1965. In the North region, the inefficient use of land contributed to lower land productivity to a great extent. In the Southwest and Central region, the high efficiency of land use relative to the national average resulted in higher land productivity. In the Northwest which has lower inputs of labor and chemical fertilizer, land was used extensively. However, the relatively higher efficiency there offset the low land productivity to some extent. In the South and Southwest, efficiency affecting land productivity was slightly above the national average.

The accounting for land productivity differences in 1986 is shown in Table 5.6. In relatively abundant land regions (Northwest and Northeast), the low land productivity relative to the national average is explained by low labor input per unit of land; in relatively land scarce regions (South and Central regions), higher land productivity (higher than national average) is explained by higher labor input to the limited land. Due to the rapid labor transfer from agriculture to rural industry in the Southeast, higher labor productivity was contributed by decreasing labor input. Power input per unit of land

Table 5.5 Accounting for Land Productivity Differences, 1965 (Regions vs National Average)

	N.E.	N.	N.W.	C.	S.E.	S.W.	S.
Difference in Land							
Productivity(%)	-21.3	-31.6	-42.2	12.6	16.0	24.4	29.3
	(100)	(100)	(100)	(100)	(100)	(100)	(100)
Labor	-49.7	-.9	-70.8	0	7.1	14.0	11.8
	(233.0)	(2.8)	(168.0)	(0)	(44.4)	(57.4)	(40.3)
C.Fertilizer	-17.7	-8.7	-46.0	1.4	4.7	-3.5	9.9
	(83.1)	(27.5)	(109)	(11.1)	(29.4)	(14.3)	(33.8)
M. Fertilizer	-10.5	-7.9	10.1	-3.6	-4.2	8.3	3.7
	(49.3)	(25.0)	(239.0)	(-28.7)	(26.3)	(34.0)	(12.6)
Power	-.27	.0	1.6	-.8	-4.9	1.2	1.7
	(1.3)	(0)	(-3.8)	(-6.3)	(-30.6)	(4.9)	(5.8)
Efficiency	56.9	-14.1	62.9	15.6	18.7	4.4	2.2
	(267.0)	(44.6)	(-149.1)	(123.8)	(116.8)	(18.1)	(7.5)

1. Accounting formula is similar to that of labor productivity.

had little effects on land productivity differences among regions in 1986 but compared to 1965, its contribution increases in explaining the differences. Chemical fertilizer is important in explaining the differences in land productivity in 1986. More than 50% of land productivity differences in the Northwest region was accounted for by the low application level of chemical fertilizer; In the Southeast and South, on the other hand, the higher land productivity came partially from the higher application of fertilizer per unit of land. Manurial fertilizer had small effects on the land productivity growth, but the differences in manurial fertilizer use per unit of land explained large shares of the land productivity differences.

The Southeast and South regions had very high efficiency in land productivity relative to the national average. Although inputs per unit of land were higher than in other regions, the efficiency improvement certainly contributed to land productivity greatly. With much lower inputs per unit of land in the Northeast, the relative higher efficiency counteracted the low land productivity to a great extent. The North region still had the relative lowest efficiency; at the same time, the Southwest region changed from being a more efficient region than national average to becoming a less efficient region.

Comparing Tables 5.5 and 5.6, we find that differences widened in land productivity among the regions. Although the effects of labor input

Table 5.6 Accounting for Land Productivity Differences, 1986
(Regions vs. National Average)

	N.E.	N.	N.W.	C.	S.E.	S.W.	S.
Difference in Land							
Productivity(%)	-23.5	-17.0	-79.2	4.1	25.1	-.4	34.5
	(100)	(100)	(100)	(100)	(100)	(100)	(100)
Labor	-25.4	-1.1	-27.1	.4	1.1	7.0	5.9
	(108.1)	(6.5)	(34.2)	(9.8)	(4.4)	(1750)	(17.1)
C. Fertilizer	-5.0	-.7	-50.3	.2	6.8	-1.6	7.6
	(21.3)	(4.1)	(63.5)	(4.9)	(27.1)	(400)	(22.0)
M.Fertilizer	-14.7	-5.1	10.1	-4.4	-8.4	7.1	2.9
	(62.5)	(.3)	(12.8)	(-107.3)	(33.5)	(1775)	(8.4)
Power	-2.6	5.4	-8.6	-5.8	.6	-10.0	2.4
	(11.1)	(31.8)	(10.9)	(-141)	(2.4)	(2500)	(7.0)
Efficiency	24.2	-15.5	-3.3	5.5	25.0	-2.9	15.7
	(103.0)	(91.2)	(4.2)	(134.1)	(99.6)	(725.0)	(45.5)

differentials have declined in accounting for land productivity, they are still important factors in the pattern of land productivity. In both 1965 and 1986, the difference in power use per unit of land had very little effect on the difference in land productivity whereas efficiency affected land productivity greatly. However, the differences among regions have grown over the period.

5.4 Accounting for Total Production Growth

In Chapter 4, I constructed a total factor productivity index; it does not, however, explain the sources of total production growth. Accounting for growth of total production is summarized in Table 5.7.

For the country as a whole, total production increased 63.9%, of which 29.6% is explained by technical change and efficiency improvement and 70.4%, by increased use of total inputs. Among all inputs, the increased use of chemical fertilizer contributed the most to production growth. The second important factor in the increase of total production from 1965 to 1986 was increased power input. Both labor and manurial fertilizer are still important in accounting for production growth, however. Land has the least effect on production growth.

The total agricultural production change varies from 51.3% in the Northwest to 68.9% in the Southeast region. However, sources of growth among the regions vary. Southeast reached the highest total agricultural production through increasing modern input usage (fertilizer and machinery), improving efficiency, and

Table 5.7 Accounting for Growth of Total Agricultural Production, 1965 to 1986

	N.E.	N.	N.W.	C.	S.E.	S.W.	S.	National
Total Production Change(%)								
	63.7	66.5	51.3	59.9	68.9	56.6	66.7	63.9
	(100)	(100)	(100)	(100)	(100)	(100)	(100)	(100)
Total Input Change(%)								
	42.2	43.8	53.2	43.7	42.2	51.3	43.2	45.0
	(66.2)	(65.9)	(103.7)	(72.9)	(61.2)	(90.6)	(64.8)	(70.4)
Labor	4.3	3.9	7.2	6.2	3.6	9.7	7.2	5.8
	(6.8)	(5.9)	(14.0)	(10.4)	(5.2)	(17.1)	(10.9)	(9.1)
Land	.5	-.8	-1.4	-0	.9	2.2	.3	.1
	(.8)	(1.2)	(-2.7)	(0)	(1.3)	(3.9)	(.5)	(.2)
C. Fertilizer	19.7	19.4	20.5	18.4	18.5	19.2	15.9	18.7
	(30.9)	(29.2)	(40.0)	(30.7)	(26.9)	(33.9)	(23.8)	(29.3)
M. Fertilizer	3.6	6.7	10.9	5.2	3.8	7.4	6.1	6.2
	(5.7)	(10.1)	(21.2)	(8.7)	(5.5)	(13.1)	(9.1)	(9.7)
Power	14.1	14.6	16.0	13.9	15.4	12.8	13.7	14.2
	(22.1)	(22.0)	(31.2)	(23.2)	(22.4)	(22.6)	(20.5)	(22.2)
Total Productivity Change(%)								
	21.5	22.7	-1.9	16.2	26.7	5.3	23.5	18.9
	(33.8)	(34.1)	(-3.7)	(27.0)	(38.8)	(9.4)	(35.2)	(29.5)
Efficiency	-3.6	10.3	8.1	11.5	11.5	2.0	26.3	10.2
	(-5.7)	(15.5)	(15.8)	(19.2)	(16.7)	(3.6)	(39.4)	(16.0)
Technology	25.1	12.4	-10.0	4.7	15.2	3.3	-2.8	8.7
	(39.4)	(18.6)	(-19.5)	(7.8)	(22.0)	(5.8)	(-4.2)	(13.6)

1. (5.3) is employed for this accounting.

making technical changes. Also, South region reached its relative high production growth mainly through increasing fertilizer and power inputs and adopting technical changes. The lower production growth in the Northeast, Northwest, and Southwest regions are explained by both low efficiency improvement and low inputs.

5.5 Sources of Technical Change and the Induced Innovation Process in Socialist Countries

We have noted earlier that technical change contributes not only to partial factor productivity growth but, also, to total factor productivity. The question is, where did the technical change come from and who did it?

Hayami and Ruttan defined technological change as any change in production coefficients resulting from a purposeful resource-using activity directed to the development of new knowledge which is embodied in design, materials, or organization (Hayami and Ruttan, 1971/1985). They treat technical change as an endogenous variable in economic growth. In terms of their definition, it is entirely rational for competitive firms to allocate funds to develop technology that will facilitate the substitution of increasingly more expensive factors for less expensive factors.

Hayami and Ruttan also explain the mechanism of induced innovation in the public sector. They hypothesize that technical change is guided along an efficient path by price signals in the market, provided that the price efficiently reflects changes in the demand and supply of products and factors and that there exists effective interaction among farmers, public research institutes, and private agricultural supply firms. The perfect market is very important to Hayami and Ruttan's hypothesis.

Here, however, we argue that even in a non-market economy such as China's the induced process still exists (Wilkin also discussed the induced innovation process in a socialist country in 1987). In the socialist country, the objective of production is to supply, within the limits of fixed resources, the maximum amount of products to meet consumers' demand. Lacking perfect price information, the government allocates resources by equalizing the marginal product of each input among regions. Thus, the roles of different sectors are as follows:

(1) Government. Government, in this process, coordinates the work of all sectors. After observing the information, such as marginal product of each input, international prices of each input, relative ratio change in input, and demand for products, government allocates research funds or investments to agriculture and other sectors to find ways to substitute for scarce resources. In allocating output targets and resources among regions, the government equalizes marginal products among regions in order to fully use regional regional advantages.

(2) Research Institute. The role of the research institute is first, to accept government advice, and second, to allocate research funds in order to discover or

invent new technology that uses abundant resources as substitutes for scarce resources.

(3) Farmers. After analyzing the government plan and the technology and resources that are available, farmers choose an appropriate technology and they allocate resources to maximize profits. They press the public research institute to develop new technology and also demand that industry supply modern inputs which can substitute for scarce inputs. Figure 5.2 shows induced innovation technical change in a non-market economy.

P_1 represents the common production function (envelopes of regional production functions, or production functions at the different time) at time 1 and P_2 represents the common production function at time 2. At time 1, for a region with factor ratio R_1, the optimal point is A. Owing to population growth or urban expansion, the factor ratio is changed from R_1 to R_2 at time 2. However, the optimal point for the region is C rather than B at time 2. The change from A to B results from factor substitution in the absence of technological change. The government observes the factor ratio change and allocates funds to invent a new technology that will substitute abundant resources for scarce resources. When the new common production becomes available, the region's production will be at point C. The change from B to C is due to the technical change.

Since 1949, the Chinese government has put great effort into encouraging the invention of new technology to relieve the land constraint on agricultural production.

Figure 5.2 Technical Change Induced by Factor Ratio Change

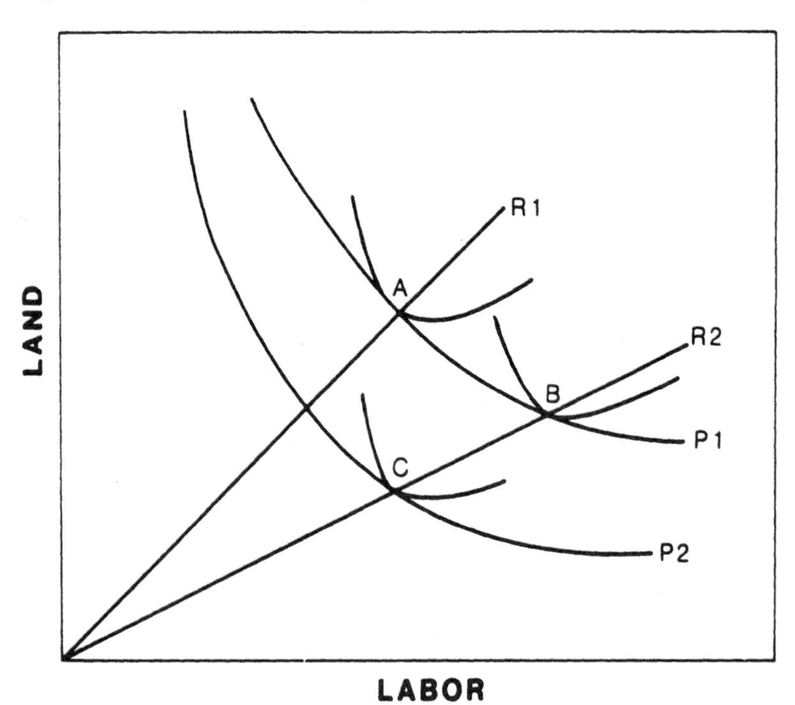

In 1956, Chinese scientists initiated the breeding program that led to the first high-yielding dwarf indica varieties of rice developed in China. These varieties had high yield potential, were responsive to fertilizer, and were relatively resistant to lodging and disease (Hsu, 1982). With adequate fertilizer and water, the farmers produced yields of 5-6 metric tons per hectare, comparable to those of the IR-8 dwarf rice developed at the International Rice Research Institute in the Philippines. The Chinese varieties, however, had a shorter growing period of 110–115 days, making it possible to expand the double cropping of rice in the South and Southeast. By 1977, these varieties were grown in more than 80% of China's total rice land (Hsu, 1982).

The breeding of wheat and other crops began in 1960s. Although the results were not as significant as those for rice, the extension of these breeding varieties has contributed greatly to wheat production in the North.

All these successes were induced by changes in land/labor ratio or other factor ratio changes. However, any imperfect information and wrong decision by the government will bias the direction of this induced innovation process.

5.6 Summary

The accounting of sources of growth and differences among regions for labor productivity, land productivity, and total factor productivity in China permits some conclusions to be drawn.

Chemical fertilizer plays a greater role than power input in land and labor productivity growth. The greater use of chemical fertilizer input increased both land and labor productivities. The increased use of power input increases labor productivity only if labor input is decreased as a result. Unfortunately, the opportunities for farmers to leave agriculture are so few that the absolute quantity of labor input increased from 1965 to 1986.

For the country as a whole, 5% of labor productivity growth is offset by the decrease of land endowment per labor unit. The decrease of absolute land input and land input per unit of labor has become a major constraint to the further increase of production.

The increased manurial fertilizer was found to contribute to both labor and land productivity growth, although the importance of its role is decreasing.

Technical change and efficiency improvement are very important to both land and labor productivity growth. They account for about 46.4% of labor productivity growth. The rest of labor productivity growth is mainly accounted for by increased use of chemical fertilizer and power input. Technical change and efficiency improvement explain about 30% of land productivity. More than 70% of land productivity growth is explained mainly by increases in labor, chemical fertilizer, and power inputs.

From the accounting of differences in labor productivity among the regions, we find that traditional inputs, such as land and manurial fertilizer explained large shares in 1965. Modern inputs--chemical fertilizer and power inputs--were not significant then. In 1986, however, the differences in land and manurial fertilizer input per unit of labor were still important but less so than earlier. Differences in chemical fertilizer and power inputs have become the main sources of labor productivity differences. In both 1965 and 1985, differences in efficiency played a crucial role in regional differences in labor productivity.

Like labor productivity differences, land productivity differences in 1965 are mainly explained by the differences in the traditional inputs of labor and manurial fertilizer. The differences in chemical fertilizer and power input had little effect on land productivity differences in 1965. As more modern inputs were used, their contribution to explaining land productivity differences has increased.

Large efficiency differences were found in land productivity. The lowest efficiency was found in the North region in both 1965 and 1986.

From the accounting of total production growth, it is evident that the traditional inputs of land, labor, and manurial fertilizer explained only a small part of total production growth. The increase of chemical fertilizer and power input accounts for more than 50% of the growth (chemical fertilizer, 29.3%; power input, 22.2%). Thus, more than 70% of total production growth is explained by the increase of total input.

Technical change and efficiency improvement accounts for 30% of total production growth. This technical change was biased to labor-using and land-saving, and it was induced by the change in land/labor ratio. Government allocated investment and research funds to find substitutes for scarce resources. The research institutes focused on projects that might lead to the discovery or invention of new technology in order to achieve this target. Farmers choose appropriate technologies to use their resources efficiently. In this process, the marginal products and international prices of resources are important signals rather than distorted prices for decision makers.

CHAPTER 6
CONCLUSIONS AND POLICY IMPLICATIONS

6.1 Introduction

Agriculture in China is being transformed from a traditional to a modern pattern. More modern inputs have been applied to attain the high growth rate of total production in order to satisfy the increasing demand. As a result, total production grew at 5% a year from 1965 to 1986, which is the most rapid among all socialist countries and is even more rapid than most of the developing countries during the same period. But questions on the efficiency of this performance had no satisfactory answers and the regional differences of this performance were not well-understood. This book fills this gap in our knowledge.

6.2 Major Conclusions

1. Comparisons of the natural resources endowment and economic conditions among regions show that the differences among them are large. These endowment and economic differences, consequently, are reflected in total

production and partial and total factor productivity growth patterns. The land/labor ratio is worsening, owing to the land and population characteristics, thus it has determined the productivity growth pattern and the agricultural technology employed. However, the differences in the patterns and adoptions of different technologies among the regions are not small.

Of all regions, Northeast and Northwest are relatively land abundant; consequently, they are more land-using and labor-saving oriented. More draft animals and machinery are applied and land is used extensively. In contrast, South, Southeast, Central, North, and Southwest regions are extremely land-scarce. The productivity pattern is characterized by the much higher growth of land than of labor productivity. Labor intensive farming is practiced with high fertilizer and other current inputs. Government policies, however, also had very strong effects on agricultural development. The grain self-sufficiency policy for each region was achieved with the loss of comparative regional advantage. The preferential policy in the allocation of modern inputs also affected regional agricultural development.

2. The estimates of the frontier productions for China's agriculture show that the coefficients of traditional inputs are not small, which indicates that the main inputs in China's agriculture are still the traditional. However, the coefficients of traditional inputs of land, labor, and manurial fertilizer are decreasing rapidly. The coefficients of modern inputs,

such as chemical fertilizer and power inputs, were small in 1965 but they increased fast. By 1986, the modern inputs were as important as traditional inputs.

Average efficiency measurement indicates that the Northeast region had the highest efficiency in agricultural production, but efficiency in the region decreased over time. In the South, Southeast and Central regions, the efficiency was similar to the national average whereas Northwest and Southwest regions had the lowest efficiency. Nevertheless the least efficient region in 1965 has shown significant improvement.

3. For the country as a whole, labor productivity grew at 4% a year from 1965 to 1986. However, the growth rates for each region are not the same. In the Southeast region, the growth rate was 5.1% whereas in the Southwest it was only 2.23% a year for the period. These figures may conceal the facts if we do not separate the whole period into two subperiods: 1965 to 1979 and 1980 to 1986.

Before 1979, owing to the adherence of labor to agriculture, labor productivity grew slowly. High growth rates in the Northeast region were achieved by the expansion of land areas. Although land productivity grew fast in Southeast, South, North, Southwest, and Central regions, labor productivity was stagnant between 1965 and 1979 as the land/labor ratio deteriorated.

After 1980, the expansion of land areas in the Northeast region was limited; at the same time, rural industry was less well-developed than in other regions and labor productivity growth rates started to decline. The emergence of rural industry after 1980 in Southeast and South regions absorbed a large quantity of labor from agriculture and thereby improved the land/labor ratio. In association with the higher growth of land productivity, labor productivity improved greatly.

Before measuring total factor productivity, the disequilibria of different factor inputs must be discussed. The marginal product of labor is much lower than its factor share, which indicates that there exists a large disguised unemployed population in China's agriculture. This widens the difference between labor factor share and its production elasticity. Land does not have a price in China, so the cost of land is difficult to measure. Thus, the residual (net profit from operation of unit land) is used as the proxy for land cost. Like the labor share, land factor share is much larger than its marginal product. The factor shares of chemical fertilizer and power input are much less than their marginal products, however.

Three different weights are used to aggregate total input. After the comparison, the most appropriate measure is chosen. All regions except Northwest had positive growth rates from 1965 to 1986. Even before 1979, which was considered the less efficient period in agricultural production, most regions improved their total factor productivity slightly. The Northwest regions showed deterioration of total factor productivity between 1965 and 1979 mainly owing to their slow total production growth. The slow growth rate of total productivity in this period can be explained by two reasons: (1) China entered the period of transformation from traditional to modern agriculture after the 1960s, hence the increase of modern inputs has been faster than the increase of total production; and (2) political movements motivated agricultural production, which resulted in the inefficient use of resources.

4. Modern inputs play important roles in explaining partial productivity (labor and land productivities) growth. The traditional inputs are still important, but the importance of its role has declined.

Differences in traditional inputs explain the big shares in the differences of partial productivities in 1965. But in 1986, the differences in modern inputs account for most of the regional differences in partial productivities. Differences in chemical fertilizer input not only affect the differences in land productivity but, also, those of labor productivity.

Efficiency improvement and technical change explain 30% to 50% of partial productivity growth. Furthermore, differences in efficiency improvement explain a large share of the differences in partial productivities among regions.

The increase of traditional inputs still explains a certain share of total production growth. Among all inputs, the increase of chemical fertilizer use is the most important source of production growth and the second important source is the increase of power input. All inputs explain about 70% of total production growth. The residual, the proxy for technical change and efficiency improvement, accounts for about 30% of total production growth.

Technical change in Chinese agriculture is biased toward land-saving and labor-using. The change in factor ratio rather than in prices induces the biased technical change.

6.3 Some Policy Implications

China's population reached 1065.29 million in 1987. The growth rate in population over the last 37 years (1949 to 1987) was 1.84%. Over the last decade, the growth rate declined (1.29% a year). Further decreases in population growth will not be easy in the next decade because the base population is large and youths born in the 1960s begin to enter the reproduction age. The demand for food will continue to grow even without the income effects. As industrialization develops, the demand for cash crops is increasing. How to meet the demand for fast increases in food and industrial material in the future is the urgent problem facing the Chinese government.

The first and quick solution for China is to increase more inputs, such as land, labor, chemical fertilizer, machinery, and so on. However, the potential for the increase of land input is very limited. It is conservatively estimated that there remains only 13 million hectares of new land in the country for crop production. This acreage is in the more remote areas of Heilongjiang province and the Xinjiang Autonomous Region.

If all these lands are reclaimed, the maximum effective cropping will most likely not exceed 10 million hectares (see Sun Han's article in Wittwer's book, 1987). Nevertheless, the total of the potential erosion of soil in China is about 1.2 million square kilometers. Since the mid-1960s, 3 million hectares have been reduced to desert and degeneration has claimed 23% of the total grassland areas. Coupled with the expansion of cities and industries, more land (especially of high quality) will have to be given up to non-agricultural activities. Thus it is likely that in the future total land input will be decreased.

Without an increase in land areas, the use of more labor will not increase total production, and the marginal return of labor will be zero. The increase of modern inputs, especially chemical fertilizer, is the major possibility for increasing total production. Although fertilizer input per unit of land in China is higher than in most developing countries, the potential output increase through the use of more fertilizer is large in some regions. They include Northeast, Northwest, North, and Southwest. Increased machinery input will not have great effects on production increase.

Thirty percent of total production growth in China is accounted for by technical change and efficiency improvement. Because of the limited land input and the diminishing returns of labor, fertilizer, and other inputs, the increasing demand for agricultural products will have to be met mainly by technical change and efficiency improvement.

Compared to other countries, the contribution of technical change to agricultural production growth in China is very small. In Japan, from 1880 to 1965, about 74% of total production growth stemmed from technical change (Hayami, 1981). In the United States, technical change accounted for 47% of the growth in U.S. total output from 1945 to 1965 (Jorgenson and Griliiches, 1966).

Underinvestment in agriculture may explain the slow technical change. Table 6.1 presents the shares of agricultural output in total output, shares of national agricultural income in total national income, and the share of agricultural investment in total investment. In 1985, the agricultural sector produced 28.1% of total output and 41.1% of national income although the agricultural investment was only 3.4% of total investment. The underinvestment of agriculture has resulted in poor rural infrastructure and insufficient agricultural research; they will be the bottleneck to the further increase of agricultural production. Nevertheless, investment in agriculture shows great regional preferences. The government allocates limited investment for political or military reasons, thus the border regions, such as Northwest and Northeast, have been favored. Table 6.2 presents the regional shares of total production and investment.

Despite the preferential investment, the differences in total production and productivity growth have widened

Table 6.1 Shares of Agriculture in Total Output and Investment(%)

	Shares of Total Output	Shares of National Income	Shares of Investment
1949	58.5	68.4	7.1
1965	30.9	46.2	17.7
1970	27.8	41.3	9.8
1975	25.0	39.4	10.5
1985	28.1	41.1	3.4

Source: <u>China's Statistical Yearbook, 1987.</u>

among the regions. In the trade-off between efficiency and equity, the Chinese government pays more attention to equity.

More efforts should be made to bring about new technologies for agriculture. China has a wide range of soils, climates, hours of daylight and other local factors. An equally wide range of crop varieties most suited to different regional and local conditions must be developed. This should be done not only for rice, wheat, and maize but, also, for sorghum, millet, potatoes, rye, oats, buckwheat, and so on. Research is also important to upgrade the breeding and care of animals, poultry, and pond-reared fish in order to maximize output from marginal land. Finally, research and development can lead to

Table 6.2 Regional Shares of Total Agricultural Production and Investment(%), 1986

	Production	National income	Investment
Northeast	8.9	9.9	10.2
North	29.4	29.2	28.8
Northwest	3.7	4.1	11.4
Central	12.5	13.9	10.0
Southeast	22.8	19.3	11.0
Southwest	11.6	11.9	8.9
South	10.7	12.5	19.4

Source: China's Statistical Yearbook, 1987.

reduced crop losses from insect pests and plant diseases through biological controls, pesticides, and herbicides (Wiens, 1982).

The recent institutional change has improved efficiency in agriculture greatly. However, the new strategy should be focused on the improvement of regional specifications, based on the regional comparative advantage. Crops should be grown where soil and climate provide the most favorable conditions. Although rural labor have more opportunities to work outside agriculture, the absence of labor mobility will become a serious source of inefficiency. The recent introduction of markets in agriculture has brought about the more efficient allocation of resources. Nevertheless, the instability of input and output prices and the insufficient supply of modern inputs will continue to affect agriculture to a great extent.

Appendix A Data Explanation

Gross Agricultural Production Value

The monetary value of the total output is estimated on the basis of production from plantations, forestry, livestock husbandry, sidelines, and fisheries. The substance of each output is as follows:

(1) Crop Plantation: grain, cotton, oilseeds, sugar, hemp, tobacco, vegetables, medical materials, melons and the activities of tea gardens, mulberry fields, and orchards.

(2) Forestry: plantations of trees and forests (excludes tea gardens, mulberry fields, and orchards), gathering of forestry products and cuts of bamboo and tree harvests by villages, production teams, and individuals.

(3) Animal husbandry: all animals rather than fish.

(4) Sidelines: gathering wild plants, hunting ,and rural industries operated by peasant families.

(5) Fisheries: raising and catching water plants and animals.

Gross agricultural production value includes all value produced by state owned specialized farms (crop production, forestry, fishery), agricultural experimental farms (institutes); collective-owned crop production, forests, animal husbandry, sidelines, and fisheries; and peasant-owned crop plantations and animal husbandry.

Gross agricultural production value is the sum of the

values from all listed activities. The value for each activity is calculated by multiplying output by its price.

Gross Rural Industry Value

It is the total values produced by township-run industries, village-run industries and the industries below village (production teams and peasant families). It is calculated by final products times their prices.

Irrigated Area

It is effective area which has water sources and the set of irrigation equipment and can be irrigated under the normal year.

Total Agricultural Machinery Power

It includes all power engaged in agricultural tillage, irrigation, harvest, agricultural product processing, transportation, plant protection, forestry and fisheries. But it excludes the power used for rural industries and non-agricultural transportation.

Sown Area

Sown area are used as land input in this study. Since land is often sown several times a year, sown area is substantially larger than cultivated area. Area of crops destroyed prior to the scheduled reporting date is excluded, but area suffering disasters after the scheduled reporting date is included. Seedbed area is not included unless it is planted with another crop after its seedlings have been transplanted. Area of crops cultivated along roadsides is excluded, but area intercropped in orchards. Area of perennial crops is included in the year the crop yields a harvest, regardless of when it was actually planted. Crops regrown from the previous season's stubble are not counted again in the second year's sown area.

Chemical Fertilizer

It has three different production units, i.e., actual, standard, and the effective weights. Actual weight is the gross weight of the product. Standard weight is the weight converted into weight of standard fertilizers: ammonium sulfate (20% nitrogen), super phosphate (18.7% P_2O_5) and potassium sulfate (40%). Effective weight measures the actual nutrient content.

Manurial Fertilizer

It is estimated from agricultural population and the numbers of livestock. The FAO estimated that one animal (horse unit) produces about 4 tons of manure per year and a person produces .25 ton per year. Manure contains 2.2% pure nutrient, and the manure availability is about 75%. Therefore, manurial resources are estimated as follows:

Manurial resources in a year (tons) =
((.25 × rural population + 4 × numbers of livestock) × 2.2%) × 75%.

This estimation underestimates the true value, since it does not include green manure. However, if we assume green manure is proportional to population and numbers of animals, it will not affect the production function coefficients and the measurement of total productivity. This estimation is not significantly different from that by Stone (1981)

Labor Force

State Statistical Bureau defines labor resources as the population from 16-59 for male and 16-54 for female. However, labor engaged in agriculture also include the population outside this range who frequently engage in agricultural production for more than three months a year. It excludes the soldiers in the army and criminals in custody and handicapped persons who lose working abilities.

Draft Animals

It includes water buffalos, cattles, horses, asses, mules and camels.

Appendix B Total Production Growth for Each Province

Table B.1 Total Production Growth for Each Province
(Annual Growth Rate %)

	1965-79	1979-86	1965-86
NORTHEAST	4.7	5.5	4.9
Liaoning	4.6	5.8	5.2
Jilin	3.9	5.7	4.6
Heilongjiang	5.1	2.9	4.9
NORTH	4.6	6.8	5.3
Beijing	3.4	7.7	4.7
Tianjin	6.0	11.7	8.5
Shanxi	4.1	5.8	4.3
Hebei	4.2	2.2	3.2
Shandong	5.1	7.8	6.5
Henan	5.8	5.6	6.1
Shaanxi	3.2	6.7	3.7
Gansu	3.2	6.9	4.4
NORTHWEST	1.6	7.3	3.5
Nei Monggol	1.9	10.9	3.5
Tibet	7.9	4.6	6.8
Qinghai	1.4	4.4	2.8
Ningxia	4.1	7.6	5.6
Xinjiang	.1	8.8	2.9
CENTRAL	4.0	5.3	4.4
Jiangxi	3.4	5.3	4.0
Hubei	3.8	8.5	4.3
Hunan	4.6	7.2	4.9
SOUTHEAST	3.3	7.8	5.7
Shanghai	5.4	5.5	5.4
Jiangsu	5.7	9.4	6.4
Zhejiang	4.8	7.3	5.2
Anhui	3.1	11.0	5.3

(Continued)

(Continued, Table B.1)

	1965-79	1979-86	1965-86
SOUTHWEST	2.9	6.4	4.0
Sichuan	3.3	5.1	4.3
Guizhou	1.4	8.3	3.4
Yunnan	2.9	5.5	3.8
SOUTH	3.3	9.6	5.4
Fujian	1.5	9.1	3.8
Guangdong	3.5	14.3	6.2
Guangxi	4.7	5.6	5.1
NATIONAL	4.0	6.9	5.0

Appendix C Provincial Labor Productivity, Land Productivity and Total Factor Productivity

Table C.1 Labor Productivity for Different Provinces (1980 Yuan per labor)

Year	1965	1970	1975	1976	1977	1978	1979
NORTHEAST	785.3	776.0	998.0	889.0	905.1	1169.6	1463.3
Liaoning	572.7	535.4	700.1	643.4	650.7	791.5	1254.7
Jilin	873.1	895.5	1065.2	952.4	968.7	1261.9	1595.8
Heilongjiang	1080.9	1109.7	1423.6	1229.6	1256.0	1683.7	1615.3
NORTH	340.7	356.9	478.3	445.9	450.1	551.3	613.3
Beijing	517.3	487.9	710.1	680.5	656.2	861.5	985.0
Tianjin	222.2	253.4	395.2	364.9	346.1	521.3	686.5
Shanxi	336.1	348.5	466.8	410.7	448.7	587.7	626.6
Hebei	403.0	399.8	590.6	480.4	504.9	574.2	694.7
Shandong	328.3	346.6	472.0	457.9	454.1	537.3	662.3
Henan	284.3	354.0	435.4	426.2	414.9	511.9	529.3
Shaanxi	444.6	377.4	490.4	458.6	457.6	552.8	630.7
Gansu	395.7	347.8	530.8	470.9	472.3	552.6	512.8
NORTHWEST	823.0	722.3	811.5	755.8	703.4	796.9	803.3
Nei Monggol	800.0	813.6	1023.9	949.3	841.8	854.6	873.2
Tibet	287.5	461.8	490.2	455.6	443.9	551.4	673.6
Qinghai	808.9	615.0	752.2	702.9	685.0	777.4	728.0
Ningxia	475.0	476.5	583.1	495.0	513.3	600.0	636.3
Xinjiang	1103.1	766.9	719.2	687.0	659.5	839.7	816.4
CENTRAL	503.3	503.2	589.3	557.8	561.5	649.5	764.1
Jiangxi	632.9	650.2	636.1	568.9	587.0	672.0	772.4
Hubei	566.4	539.1	645.7	636.9	635.2	751.6	863.2
Hunan	398.6	411.1	524.8	492.1	493.3	560.8	681.7
SOUTHEAST	436.2	438.9	506.7	515.0	488.3	617.4	738.9
Shanghai	380.4	361.3	554.6	574.3	546.8	741.5	952.9
Jiangsu	366.2	374.7	458.6	486.4	463.3	605.5	828.6
Zhejiang	547.2	534.3	595.6	565.9	573.9	773.4	783.8
Anhui	474.0	476.7	505.3	509.9	454.4	506.2	566.5

(Continued)

(Continued, Table C.1)

Year	1965	1970	1975	1976	1977	1978	1979
SOUTHWEST	390.2	325.0	341.0	321.0	346.5	438.6	474.6
Sichuan	379.6	318.2	313.4	301.2	339.2	437.7	500.4
Guizhou	469.2	362.5	370.4	351.3	374.4	430.8	427.5
Yunnan	366.7	319.4	409.5	361.8	348.2	447.6	432.4
SOUTH	449.7	444.4	443.2	482.3	524.0	626.6	599.7
Fujian	707.1	665.1	665.3	602.0	646.9	840.6	738.3
Guangdong	434.0	424.9	427.6	533.9	578.5	632.9	613.2
Guangxi	333.3	350.8	341.5	332.6	367.9	494.2	507.3
National	439.6	432.2	510.1	490.6	497.1	609.4	673.3

(Continued)

(Continued, Table C.1)

Year	1980	1981	1982	1983	1984	1985	1986
NORTHEAST	1608.7	1620.3	1665.8	1859.1	2009.5	1879.9	1870.3
Liaoning	1211.3	1224.5	1237.0	1392.7	1514.0	1485.5	1669.5
Jilin	1611.4	1678.5	1725.6	1904.8	2128.7	1846.6	1613.7
Heilongjian	2196.1	2177.7	2289.2	2552.6	2643.4	2457.3	2407.0
NORTH	614.8	629.0	676.8	764.4	868.5	941.8	897.5
Beijing	941.9	964.7	1045.2	1204.1	1429.6	2179.8	2445.3
Tianjin	781.4	589.6	703.4	701.7	908.8	1474.8	2211.6
Shanxi	570.2	587.2	666.9	756.8	845.1	914.6	797.4
Hebei	610.0	622.9	751.6	761.7	948.5	978.8	784.0
Shandong	704.1	726.4	781.7	895.6	1070.0	1169.1	1155.9
Henan	571.5	608.9	580.4	705.2	747.9	782.1	731.3
Shaanxi	547.4	543.6	650.1	640.6	716.9	771.1	764.1
Gansu	517.2	476.4	533.9	590.7	628.1	737.9	743.4
NORTHWEST	287.1	979.8	1052.4	1080.9	1205.2	1350.9	1338.2
Nei Monggol	790.2	915.3	1015.0	1057.3	1181.1	1314.6	1214.6
Tibet	714.1	762.8	742.5	621.8	757.1	586.6	882.6
Qinghai	818.7	734.8	765.8	773.6	848.7	966.4	1014.8
Ningxia	706.8	724.8	704.1	789.8	901.1	928.2	940.7
Xinjiang	1203.7	1380.7	1517.4	1523.0	1671.8	1918.0	1927.4
CENTRAL	682.8	723.3	793.3	802.3	871.8	966.8	1007.6
Jiangxi	788.4	821.0	883.7	866.9	847.0	969.1	941.4
Hubei	733.7	782.1	867.9	867.4	1069.5	1226.2	1252.8
Hunan	595.4	634.4	697.6	725.8	745.2	793.4	873.8
SOUTHEAST	677.7	815.0	824.0	827.3	959.9	1118.3	1241.5
Shanghai	818.3	907.5	1090.2	1075.3	1332.1	2063.1	2660.4
Jiangsu	757.9	807.0	919.0	966.2	1150.1	1366.4	1543.8
Zhejiang	686.7	1089.9	804.6	765.9	881.0	1031.0	1142.7
Anhui	553.8	666.8	694.4	687.6	771.7	876.5	952.1
SOUTHWEST	494.8	503.3	544.8	572.9	627.0	650.7	620.2
Sichuan	536.1	539.4	580.3	617.7	670.1	699.4	664.3
Guizhou	411.8	422.5	487.4	487.5	531.7	523.1	582.6
Yunnan	432.9	457.0	483.1	504.8	574.8	611.1	521.2

(Continued, Table C.1)

Year	1980	1981	1982	1983	1984	1985	1986
SOUTH	576.2	597.1	672.9	687.2	740.2	846.2	1040.8
Fujian	707.6	731.5	781.7	808.6	916.9	1023.2	1158.0
Guangdong	567.6	590.8	686.7	717.1	809.5	1002.4	1358.9
Guangxi	517.6	534.9	597.2	584.8	563.0	573.7	613.0
NATIONAL	659.0	697.7	746.3	793.5	883.9	959.5	989.2

Table C.2 Labor Productivity Growth Rates for Each Province (% annually)

Year	1965-79	1980-86	1965-86
NORTHEAST	4.6	2.5	4.2
Liaoning	6.2	5.5	5.2
Jilin	4.8	0.0	3.0
Heilongjiang	3.1	1.5	3.9
NORTH	4.3	6.5	4.7
Beijing	5.1	17.2	7.7
Tianjin	9.1	18.9	11.6
Shanxi	4.9	5.8	4.2
Hebei	4.3	4.3	3.2
Shandong	5.6	8.6	6.2
Henan	4.9	4.2	4.6
Shaanxi	2.7	5.7	2.6
Gansu	2.0	6.2	3.1
NORTHWEST	-.17	7.1	2.3
Nei Monggol	.7	8.2	2.2
Tibet	6.8	3.6	5.5
Qinghai	-0.8	3.6	1.1
Ningxia	2.3	4.9	3.3
Xinjiang	-2.3	8.2	2.7
CENTRAL	3.0	6.7	3.4
Jiangxi	1.5	3.0	1.9
Hubei	3.3	9.3	3.9
Hunan	4.2	6.6	3.8
SOUTHEAST	3.8	10.6	5.1
Shanghai	7.3	21.7	9.7
Jiangsu	6.5	11.6	7.1
Zhejiang	2.8	8.9	3.6
Anhui	1.4	9.5	3.4
SOUTHWEST	1.4	3.9	2.2
Sichuan	2.2	3.6	2.7
Guizhou	-0.7	6.0	1.0
Yunnan	1.3	3.1	1.7

(Continued)

(Continued, Table C.2)

Year	1965-79	1980-86	1965-86
SOUTH	2.1	9.6	4.1
Fujian	0.3	8.6	2.4
Guangdong	2.7	15.7	5.6
Guangxi	3.3	2.9	2.9
NATIONAL	3.3	7.0	3.9

Table C.3 Land Productivity for Different Provinces
(Yuan Per Mu of Agricultural Land)

Year	1965	1970	1975	1976	1977	1978	1979	1980
NORTHEAST	40.8	48.5	67.6	63.8	64.2	72.9	73.3	81.5
Liaoning	52.1	58.1	82.1	82.8	83.2	89.0	107.4	119.1
Jilin	40.1	48.9	65.6	64.0	64.5	74.0	73.3	79.9
Heilongjiang	34.3	42.1	60.3	53.3	53.9	63.8	57.4	65.3
NORTH	37.6	47.8	66.6	63.6	63.7	69.1	73.0	80.6
Beijing	91.8	94.1	124.6	127.2	121.4	138.5	126.4	136.5
Tianjin	34.1	38.7	59.9	66.3	56.7	72.6	80.1	103.6
Shanxi	41.9	49.8	60.8	54.2	58.5	67.8	68.1	68.8
Hebei	34.2	41.2	59.5	53.7	56.0	56.4	63.5	61.5
Shandong	41.3	54.1	77.8	77.3	76.3	80.2	90.3	109.5
Henan	31.4	49.2	66.9	68.0	65.4	71.7	72.1	83.5
Shaanxi	37.6	40.4	58.7	53.9	53.1	56.7	64.0	60.3
Gansu	31.2	32.6	57.0	51.0	50.6	52.4	49.6	52.8
NORTHWEST	34.8	36.4	45.4	42.1	39.0	41.1	43.9	44.6
Nei Monggol	33.1	38.5	52.5	49.6	44.2	45.1	46.3	42.0
Tibet	14.1	23.5	28.3	26.6	25.8	28.4	40.3	42.3
Qinghai	42.3	39.9	56.0	51.8	50.4	51.2	51.1	57.6
Ningxia	12.4	16.0	22.7	19.1	19.8	20.6	21.8	25.2
Xinjiang	52.7	45.5	47.5	43.5	41.0	45.7	50.3	54.7
CENTRAL	56.7	66.8	81.6	77.3	76.2	77.4	92.0	90.6
Jiangxi	49.4	61.2	68.5	60.7	62.2	63.1	78.6	84.0
Hubei	62.1	70.4	85.3	85.2	83.2	86.6	100.1	91.1
Hunan	56.7	67.7	87.1	81.6	79.7	78.8	93.5	94.6
SOUTHEAST	58.9	70.2	83.8	89.8	83.4	93.7	107.4	106.9
Shanghai	22.0	90.8	134.6	145.1	137.8	163.4	176.7	106.6
Jiangsu	56.8	64.7	83.3	95.8	87.3	101.2	117.0	112.8
Zhejiang	74.5	85.4	99.2	94.8	94.9	112.8	140.5	136.4
Anhui	49.6	64.7	70.3	74.8	66.6	66.4	71.7	77.3
SOUTHWEST	65.5	66.5	73.7	69.8	74.3	82.8	87.9	98.3
Sichuan	66.5	67.4	70.2	68.0	75.3	85.2	93.7	105.5
Guizhou	73.5	73.6	75.4	74.0	78.3	79.8	80.6	88.0
Yunnan	56.7	59.3	83.0	71.8	68.6	78.2	76.6	83.9

(Continued)

(Continued, Table C.3)

Year	1965	1970	1975	1976	1977	1978	1979	1980
SOUTH	70.0	77.5	75.6	88.1	94.4	99.5	96.8	103.8
Fujian	115.5	122.2	116.3	113.8	121.0	138.8	115.4	125.5
Guangdong	68.9	74.9	75.3	101.2	108.4	104.8	103.1	105.9
Guangxi	49.0	58.4	54.8	56.4	61.3	72.2	78.6	98.5
NATIONAL	49.5	57.6	71.3	70.1	70.1	76.5	82.5	87.3

(Continued)

(Continued, Table C.3)

Year	1981	1982	1983	1984	1985	1986
NORTHEAST	83.8	89.4	107.3	115.5	106.7	110.1
Liaoning	125.8	132.4	158.8	169.3	155.0	177.9
Jilin	85.4	90.5	117.7	128.7	118.9	112.0
Heilongjiang	64.6	69.7	80.1	85.7	80.0	79.8
NORTH	84.0	94.7	106.3	119.0	123.1	116.2
Beijing	138.7	146.4	172.6	194.4	211.9	231.7
Tianjin	83.3	105.2	107.1	133.2	157.0	231.7
Shanxi	69.6	90.4	102.3	111.1	115.6	99.2
Hebei	65.4	80.7	81.1	95.5	92.3	75.5
Shandon	115.4	126.4	142.2	165.0	169.8	164.2
Henan	89.9	87.9	105.3	111.9	114.1	105.7
Shaanxi	64.9	81.9	81.9	94.2	97.5	96.4
Gansu	50.3	56.8	63.1	67.5	77.0	78.4
NORTHWEST	51.2	56.5	60.8	68.5	76.7	75.9
Nei Monggol	52.4	61.0	64.4	73.3	82.2	78.6
Tibet	46.4	45.9	38.2	47.6	53.6	55.8
Qinghai	53.8	57.9	59.6	64.6	72.3	74.6
Ningxi	27.4	28.4	32.5	38.3	39.5	40.8
Xinjiang	61.5	65.7	73.9	82.8	94.3	94.6
CENTRAL	98.9	110.3	114.5	128.2	137.5	141.8
Jiangxi	89.0	97.5	98.4	110.1	118.8	117.0
Hubei	102.3	112.8	113.9	137.9	147.7	151.2
Hunan	102.6	116.8	126.4	131.6	141.1	150.4
SOUTHEAST	118.1	132.0	136.1	156.5	164.6	181.7
Shanghai	178.0	209.0	194.4	227.3	219.6	253.6
Jiangsu	124.7	141.5	151.3	175.4	181.4	198.7
Zhejian	137.3	159.9	156.5	184.4	194.8	220.7
Anhui	93.7	98.4	101.9	113.9	123.3	137.0
SOUTHWEST	102.7	114.7	125.2	136.8	141.1	135.4
Sichuan	107.9	119.9	133.7	144.7	149.2	143.8
Guizhou	93.5	109.0	113.0	122.0	121.3	129.0
Yunnan	93.3	102.9	108.7	124.3	132.0	115.4

(Continued)

(Continued, Table C.3)

Year	1981	1982	1983	1984	1985	1986
SOUTH	112.4	129.9	137.7	150.3	170.2	207.8
Fujian	133.8	148.2	157.2	182.7	202.3	226.9
Guangdong	115.6	136.0	145.0	160.1	191.0	256.1
Guangxi	96.9	112.4	117.7	119.5	124.7	132.8
NATIONAL	93.4	104.6	113.4	126.3	131.9	136.0

Table C.4 Growth Rates of Land Productivity for Each Province (% annually)

	1965-79	1980-86	1965-86
NORTHEAST	4.3	5.	4.8
Liaoning	5.7	6.9	6.0
Jilin	4.8	5.8	5.0
Heilongji	4.1	3.4	4.1
NORTH	4.8	6.3	5.5.
Beijing	2.5	9.2	4.5
Tianjin	6.8	14.4	9.6
Shanxi	3.8	6.3	4.2
Hebei	4.9	3.5	3.8
Shandong	6.2	7.0	6.8
Henan	6.6	4.0	6.0
Shaanxi	4.2	8.1	4.6
Gansu	3.6	6.8	4.5
NORTHWEST	1.7	9.1	3.8
Nei Monggol	2.6	11.0	4.2
Tibet	8.4	4.8	6.8
Qinghai	1.5	4.4	2.7
Ningxia	4.5	8.3	5.9
Xinjiang	-0.4	9.6	2.8
CENTRAL	3.5	6.2	4.5
Jiangxi	3.7	5.7	4.2
Hubei	3.7	8.8	4.3
Hunan	3.9	8.0	4.8
SOUTHEAST	4.4	9.2	5.5
Shanghai	6.1	7.9	5.5
Jiangsu	5.7	9.9	6.2
Zhejiang	5.0	8.4	5.3
Anhui	2.9	10.0	5.0
SOUTHWEST	2.1	5.4	3.5
Sichuan	2.8	5.3	3.7
Guizhou	0.7	6.6	2.7
Yunnan	2.3	5.5	3.4

(Continued)

(Continued, Table C.4)

	1965-79	1980-86	1965-86
SOUTH	2.3	12.3	5.3
Fujian	0.0	10.4	3.3
Guangdong	3.1	15.9	6.5
Guangxi	3.7	6.8	4.9
NATIONAL	4.0	7.7	4.9

Table C.5 Total Factor Productivity Index (Aggregated by Production Elasticities with Draft Animal as an Input)

Year	1965	1970	1975	1976	1977	1978	1979	1980	1981
NORTHEAST	100	88.3	102.4	92.6	90.5	103.3	102.1	108.1	108.6
Liaoning	100	99.9	106.1	98.0	94.5	100.2	111.0	113.9	116.7
Jilin	100	84.2	89.7	82.3	83.7	98.5	98.8	98.1	101.9
Heilongjiang	100	82.4	111.8	97.4	92.7	109.2	97.1	111.8	108.4
NORTH	100	98.6	111.4	100.7	96.4	102.2	105.2	109.9	111.2
Beijing	100	94.1	107.1	102.3	95.6	109.9	104.5	108.0	109.5
Tianjin	100	91.0	83.4	76.3	64.1	80.5	87.4	110.0	88.8
Shanxi	100	92.7	97.8	82.3	85.0	96.0	93.2	91.0	94.2
Hebei	100	94.8	112.4	94.5	93.6	92.8	102.6	96.6	100.8
Shandong	100	101.3	121.4	112.6	104.9	106.4	116.4	130.6	130.5
Henan	100	114.5	121.1	115.6	108.1	116.1	116.1	125.1	127.8
Shaanxi	100	94.1	82.6	96.3	84.6	81.8	86.7	78.9	81.8
Gansu	100	91.0	82.0	100.0	87.8	91.0	84.4	88.5	82.2
NORTHWEST	100	92.9	95.2	85.9	76.9	79.2	79.1	79.5	88.6
Nei Monggol	100	129.7	147.8	118.8	101.1	98.8	88.7	78.4	81.2
Tibet	100	164.6	180.0	161.4	150.6	161.7	212.5	224.3	255.9
Qinghai	100	64.9	70.5	63.8	60.3	60.3	56.3	63.5	59.5
Ningxia	100	93.1	101.6	84.9	86.8	90.8	93.7	106.5	111.0
Xinjiang	100	83.7	77.1	70.0	63.8	70.1	69.4	75.8	98.1
CENTRAL	100	97.9	107.6	98.2	93.9	94.3	106.2	96.6	102.2
Jiangxi	100	103.7	114.8	98.9	97.6	111.2	111.1	115.4	118.5
Hubei	100	96.4	99.1	94.3	89.3	91.8	101.0	85.3	92.3
Hunan	100	97.4	111.2	100.4	95.3	93.6	106.5	98.0	103.0
SOUTHEAST	100	101.8	106.4	104.2	92.1	99.7	109.7	102.1	110.1
Shanghai	100	100.4	129.3	130.9	120.3	141.9	150.7	133.4	148.6
Jiangsu	100	97.8	106.8	110.0	96.1	107.5	122.3	113.1	119.1
Zhejiang	100	102.8	109.5	97.6	91.6	103.8	113.6	102.9	111.1
Anhui	100	101.8	95.0	92.2	77.9	74.7	79.1	76.2	85.8
SOUTHWEST	100	85.3	80.2	72.2	74.2	80.8	83.1	88.4	89.4
Sichuan	100	85.3	75.6	69.0	72.5	79.4	84.2	91.2	90.4
Guizhou	100	87.4	69.5	65.4	66.7	67.1	65.9	66.8	68.1
Yunnan	100	86.6	103.0	88.5	82.7	94.3	89.6	92.3	100.2

(Continued)

(Continued, Table C.5)

Year	1965	1970	1975	1976	1977	1978	1979	1980	1981
SOUTH	100	107.9	96.9	104.5	108.8	114.1	106.4	105.1	107.2
Fujian	100	96.4	91.1	80.9	82.0	91.7	73.3	72.4	73.2
Guangdong	100	105.8	99.9	123.6	128.6	123.8	116.7	110.1	112.8
Guangxi	100	125.2	91.3	88.6	94.2	111.4	115.9	123.0	125.6
NATIONAL	100	98.2	103.7	97.1	93.3	99.3	102.8	102.4	106.3

(Continued)

(Continued, Table C.5)

Year	1982	1983	1984	1985	1986
NORTHEAST	96.8	109.4	111.2	100.3	97.8
Liaoning	119.4	134.9	137.4	125.3	135.6
Jilin	102.6	123.3	126.2	111.2	98.3
Heilongjiang	112.1	123.1	125.2	115.3	111.1
NORTH	115.9	123.9	134.1	135.7	123.2
Beijing	114.2	130.9	149.6	175.8	181.6
Tianjin	107.3	102.4	124.1	148.3	206.3
Shanxi	106.7	113.4	117.4	118.8	97.7
Hebei	120.0	115.5	128.1	120.3	95.2
Shandong	134.8	145.2	168.6	170.5	157.9
Henan	115.0	130.9	132.7	132.2	120.4
Shaanxi	96.3	92.5	101.7	103.0	97.4
Gansu	87.6	94.0	95.2	105.9	99.9
NORTHWEST	89.7	90.1	97.5	104.8	94.8
Nei Monggol	88.5	86.9	93.9	97.4	85.4
Tibet	212.1	147.8	175.5	220.5	235.4
Qinghai	59.2	59.4	64.6	73.4	70.4
Ningxia	108.8	118.9	133.7	131.7	117.9
Xinjiang	85.9	89.1	94.7	104.4	96.0
CENTRAL	109.1	107.6	118.7	122.8	116.9
Jiangxi	118.5	116.0	122.4	126.6	115.9
Hubei	98.1	94.7	111.8	117.7	111.2
Hunan	112.3	113.5	120.4	122.5	120.6
SOUTHEAST	115.2	114.4	128.8	133.9	137.9
Shanghai	172.6	160.3	187.5	194.7	217.6
Jiangsu	129.6	134.2	154.1	160.7	158.4
Zhejiang	114.0	107.4	123.7	130.3	132.6
Anhui	84.1	83.1	89.4	94.0	101.5
SOUTHWEST	96.7	99.4	106.5	107.3	96.2
Sichuan	98.1	103.0	109.7	110.8	98.1
Guizhou	77.5	76.3	81.1	79.1	81.6
Yunnan	103.0	102.1	113.7	116.8	96.9

(Continued)

(Continued, Table C.5)

Year	1982	1983	1984	1985	1986
SOUTH	116.9	117.4	125.1	136.1	158.3
Fujian	75.2	76.5	85.9	92.6	95.6
Guangdong	126.5	127.2	138.7	159.6	205.5
Guangxi	136.9	135.9	133.9	132.9	135.7
NATIONAL	111.7	115.6	125.1	127.6	123.6

Table C.6 Total Factor Productivity Index (Aggregated by Production Elasticities without Draft Animal as an Input)

Year	1965	1970	1975	1976	1977	1978	1979	1980	1981
NORTHEAST	100	84.9	94.0	84.5	81.3	92.0	90.4	94.9	94.9
Liaoning	100	98.6	98.1	90.1	86.3	91.1	100.4	102.6	104.7
Jilin	100	78.9	77.7	70.2	69.7	80.7	80.4	78.8	81.4
Heilongjiang	100	78.3	103.6	90.7	84.2	98.5	87.0	99.2	95.8
NORTH	100	97.3	101.7	91.3	86.7	91.3	93.3	97.1	97.8
Beijing	100	92.8	103.1	98.1	91.4	104.8	99.8	102.4	103.6
Tianjin	100	90.9	82.8	75.6	63.3	79.3	86.0	108.0	87.2
Shanxi	100	93.0	93.5	78.3	80.5	90.5	87.5	85.2	88.0
Hebei	100	90.8	100.4	83.7	82.0	80.8	88.7	83.1	86.3
Shandong	100	101.2	109.5	100.7	92.9	93.6	101.6	113.4	112.7
Henan	100	109.8	105.6	100.0	92.6	98.6	97.8	104.8	106.6
Shaanxi	100	80.8	84.0	73.0	67.5	69.4	72.9	65.9	67.9
Gansu	100	79.5	85.2	73.7	71.6	74.6	68.6	71.6	66.2
NORTHWEST	100	97.6	113.7	105.2	96.9	103.8	107.1	110.8	127.8
NeiMonggol	100	127.5	134.3	106.6	89.8	87.2	77.2	67.8	70.3
Tibet	100	164.9	159.8	139.9	128.9	132.4	168.4	171.5	197.8
Qinghai	100	63.7	65.7	58.7	55.0	54.5	50.2	56.1	52.8
Ningxia	100	89.9	91.7	75.4	76.2	78.8	80.5	91.0	94.5
Xinjiang	100	82.5	76.1	68.5	61.9	67.1	65.7	71.4	92.2
CENTRAL	100	93.4	96.7	87.3	82.6	81.9	91.4	82.6	86.9
Jiangxi	100	97.9	104.1	88.6	86.4	84.8	96.1	95.3	98.6
Hubei	100	93.0	89.4	84.3	79.1	80.5	87.9	73.8	79.4
Hunan	100	92.4	99.1	88.4	82.8	80.3	90.3	82.5	85.9
SOUTHEAST	100	97.7	95.1	92.3	81.0	87.1	95.2	88.2	94.7
Shanghai	100	99.1	126.9	128.3	117.8	138.8	147.3	130.3	145.0
Jiangsu	100	94.3	95.8	98.0	85.1	94.6	107.0	98.4	103.1
Zhejiang	100	96.7	94.3	83.4	77.7	87.3	94.8	85.3	91.5
Anhui	100	100.2	84.2	80.6	67.3	63.9	67.1	64.4	72.3
SOUTHWEST	100	80.1	68.4	61.0	61.1	65.4	66.1	69.6	69.9
Sichuan	100	79.7	63.0	56.5	58.4	62.9	65.6	70.1	68.8
Guizhou	100	79.2	59.6	54.9	54.1	52.8	50.9	51.3	52.0
Yunnan	100	82.8	90.9	77.1	71.1	79.8	74.7	76.5	83.0

(Continued)

(Continued, Table C.6)

Year	1965	1970	1975	1976	1977	1978	1979	1980	1981
SOUTH	100	103.3	85.5	91.0	93.5	96.9	89.6	87.8	89.2
Fujian	100	88.8	79.5	69.7	69.7	77.1	60.9	59.6	59.8
Guangdong	100	103.3	90.9	111.1	114.5	109.2	102.2	95.9	97.9
Guangxi	100	119.1	75.6	72.1	75.3	87.6	90.0	94.7	96.2
NATIONAL	100	95.1	93.4	86.6	82.4	86.9	89.2	88.4	91.3

(Continued)

(Continued, Table C.6)

Year	1982	1983	1984	1985	1986
NORTHEAST	96.8	109.4	111.2	100.3	97.8
Liaoning	106.8	120.0	121.7	110.5	118.9
Jilin	81.6	97.3	99.0	86.6	75.6
Heilongjiang	98.7	107.9	109.3	100.3	96.3
NORTH	101.7	108.1	116.4	117.3	106.1
Beijing	108.0	123.4	140.6	164.4	169.6
Tianjin	105.2	100.3	121.5	145.0	201.6
Shanxi	99.6	105.4	108.8	109.8	90.0
Hebei	102.3	97.9	107.8	100.6	79.3
Shandong	115.8	124.0	143.2	144.2	132.9
Henan	95.4	108.2	109.2	108.4	98.0
Shaanxi	79.5	75.9	82.9	83.5	78.5
Gansu	70.2	74.9	75.4	83.4	78.2
NORTHWEST	80.0	80.1	86.5	92.2	82.9
Nei Monggol	76.6	75.0	80.7	83.3	72.3
Tibet	163.2	115.4	136.1	163.4	172.9
Qinghai	52.5	52.6	56.7	64.4	61.4
Ningxia	92.4	100.3	112.1	109.8	97.6
Xinjiang	80.7	83.3	88.2	96.6	88.7
CENTRAL	92.3	90.5	99.3	102.1	96.4
Jiangxi	100.8	98.1	102.6	105.4	95.8
Hubei	84.1	80.8	94.9	99.4	93.3
Hunan	93.2	93.4	98.5	99.6	97.1
SOUTHEAST	98.6	97.4	109.2	112.9	115.8
Shanghai	168.3	156.2	182.5	189.4	211.5
Jiangsu	111.7	115.1	131.4	136.4	133.8
Zhejiang	93.3	87.3	99.9	104.5	105.7
Anhui	70.6	69.5	74.3	77.9	83.4
SOUTHWEST	75.1	76.5	81.3	81.1	71.9
Sichuan	74.1	77.0	81.2	81.1	71.0
Guizhou	58.7	56.9	60.2	58.0	58.7
Yunnan	85.1	83.7	92.6	94.5	77.7

(Continued)

(Continued, Table C.6)

Year	1982	1983	1984	1985	1986
SOUTH	96.8	96.6	102.3	110.8	127.7
Fujian	61.0	61.6	68.8	73.6	75.3
Guangdong	109.3	109.3	118.8	136.1	174.2
Guangxi	104.3	103.0	100.7	99.3	100.0
NATIONAL	95.6	98.3	105.8	107.4	103.4

Table C.7 Total Factor Productivity Index
(Aggregated by Factor Shares)

Year	1965	1970	1975	1976	1977	1978	1979	1980	1981
NORTHEAST	100	101.5	127.7	116.9	117.4	138.8	145.7	157.3	159.5
Liaoning	100	118.4	140.2	131.4	129.4	138.8	151.5	160.6	167.4
Jilin	100	94.7	110.3	106.0	111.9	137.3	141.5	143.7	151.3
Heilongjiang	100	100.8	139.8	125.6	123.5	149.7	136.9	159.8	156.6
NORTH	100	108.1	138.0	128.0	126.0	139.4	147.3	154.8	158.5
Beijing	100	98.8	125.6	120.6	113.6	131.5	126.5	131.2	133.7
Tianjin	100	104.3	85.7	81.1	71.2	92.6	103.6	131.7	105.6
Shanxi	100	105.8	125.3	108.5	115.2	133.2	132.3	130.3	136.9
Hebei	100	106.8	142.5	122.9	124.6	125.9	142.2	134.5	140.4
Shandong	100	108.1	142.1	135.3	129.3	133.7	149.1	170.7	172.5
Henan	100	127.9	155.9	152.1	145.5	159.4	163.7	180.4	187.2
Shaanxi	100	92.9	121.5	111.3	108.0	115.7	126.7	115.8	120.0
Gansu	100	94.0	143.4	126.5	125.2	132.0	122.7	127.2	116.6
NORTHWEST	100	97.6	113.7	105.2	96.9	103.8	107.1	110.8	127.8
Nei Monggol	100	120.2	150.4	139.6	121.8	121.4	122.6	110.6	132.5
Tibet	100	165.4	197.5	182.0	174.6	192.3	262.7	277.8	290.8
Qinghai	100	81.5	103.1	94.8	91.0	92.3	87.3	98.7	89.3
Ningxia	100	108.3	143.0	122.8	128.7	137.7	145.2	163.7	169.7
Xinjiang	100	85.5	85.3	80.7	76.6	87.2	89.6	99.0	144.7
CENTRAL	100	105.8	123.3	115.3	113.3	118.6	137.4	127.2	136.1
Jiangxi	100	112.1	121.7	108.0	109.9	111.9	131.4	134.5	140.7
Hubei	100	103.5	122.6	119.7	116.4	122.3	137.7	119.9	131.4
Hunan	100	112.2	139.1	128.4	124.7	124.6	144.2	135.1	143.9
SOUTHEAST	100	105.9	121.1	123.0	112.4	127.6	144.7	136.1	151.3
Shanghai	100	106.0	149.6	152.8	141.9	168.5	178.9	160.2	180.6
Jiangsu	100	105.8	130.1	138.5	124.8	143.2	166.3	157.2	167.5
Zhejiang	100	108.0	122.4	113.6	111.0	129.4	146.2	136.1	146.6
Anhui	100	110.3	115.5	115.4	100.4	98.6	107.5	105.9	121.7
SOUTHWEST	100	90.7	95.1	88.7	93.6	107.2	113.5	122.1	125.2
Sichuan	100	92.0	91.5	86.4	94.0	106.1	116.5	128.8	130.1
Guizhou	100	88.1	89.4	86.4	90.7	93.4	94.4	96.2	99.2
Yunnan	100	97.7	128.4	112.4	107.2	123.9	120.1	124.4	133.9

(Continued)

(Continued, Table C.7)

Year	1965	1970	1975	1976	1977	1978	1979	1980	1981
SOUTH	100	106.1	101.8	112.4	119.6	130.0	123.5	123.7	128.7
Fujian	100	101.7	102.5	93.0	96.9	110.5	90.1	90.7	92.6
Guangdong	100	108.4	108.7	136.0	143.9	139.9	133.5	127.1	130.1
Guangxi	100	119.3	109.8	108.1	117.4	140.9	149.3	160.0	165.4
NATIONAL	100	98.2	103.7	97.1	93.3	99.3	102.8	102.4	106.3

(Continued)

(Continued, Table C.7)

Year	1982	1983	1984	1985	1986
NORTHEAST	165.1	190.1	198.1	181.7	180.4
Liaoning	173.2	199.4	205.4	187.1	203.4
Jilin	155.7	191.2	199.6	178.2	161.3
Heilongjiang	163.3	183.4	190.7	173.8	171.0
NORTH	169.2	185.5	204.3	210.9	195.4
Beijing	141.2	163.1	187.0	217.2	225.0
Tianjin	127.9	124.2	154.6	182.3	252.7
Shanxi	157.2	169.4	177.2	179.4	148.1
Hebei	168.6	164.7	189.9	177.7	142.1
Shandong	180.2	196.7	231.9	235.5	219.1
Henan	173.1	201.9	206.3	206.9	191.4
Shaanxi	144.7	142.8	162.4	165.6	155.8
Gansu	128.4	141.0	145.2	162.6	154.9
NORTHWEST	133.2	139.6	155.4	171.2	157.6
Nei Monggol	149.5	157.3	175.8	188.4	161.5
Tibet	284.3	237.9	284.1	328.5	308.7
Qinghai	93.8	93.6	100.8	112.2	106.4
Ningxia	166.7	187.4	212.8	210.6	186.6
Xinjiang	114.8	122.8	133.6	148.9	131.4
CENTRAL	147.8	148.1	163.8	173.5	171.1
Jiangxi	148.2	144.2	153.5	160.2	147.5
Hubei	143.3	141.1	169.8	180.1	170.6
Hunan	159.1	162.0	170.7	175.5	172.4
SOUTHEAST	159.0	160.4	182.4	194.6	207.4
Shanghai	211.8	198.6	233.1	237.8	268.4
Jiangsu	184.8	193.9	224.9	235.5	233.2
Zhejiang	154.3	147.4	171.8	182.4	185.5
Anhui	122.1	121.7	133.1	141.7	154.9
SOUTHWEST	136.7	144.1	156.1	159.6	147.3
Sichuan	142.2	152.4	164.0	168.4	149.8
Guizhou	113.9	113.7	123.5	119.6	125.5
Yunnan	141.7	143.4	159.8	164.6	139.3

(Continued)

(Continued, Table C.7)

Year	1982	1983	1984	1985	1986
SOUTH	143.2	146.4	157.3	174.6	207.8
Fujian	96.2	98.7	112.0	121.4	124.7
Guangdong	146.9	148.9	162.8	188.3	244.7
Guangxi	183.1	181.4	178.7	177.8	183.5
NATIONAL	111.7	115.6	125.1	127.6	123.6

Appendix D Results of Production Functions with Different Specifications and Estimation Techniques

Table D.1 Results of Production Functions from Fixed Effects and Random Effects Model

	OLS	Fixed Effects	Random Effects
Cofficients of			
Land	.318	.278	.392
	(45.6)	(10.28)	(73.45)
labor	.222	.222	.144
	(42.4)	(.006)	(35.48)
Power	.171	.109	.297
	(25.4)	(.169)	(70.93)
Chemical Fertilizer	.266	.119	.171
	(103.5)	(.296)	(76.54)
Manurial Fertilizer	.06	.194	.019
	(16.9)	(6.58)	(4.56)
Region Effects			
region 1		.024	
region 2		-.305	
region 3		-.381	
region 4		.005	
region 5		.039	
region 6		-.122	
region 7		.065	

Note: The coefficients of region 2 to 7 in column 1 are dummy variables. Observations for all regressions are 407.

Table D.2 Estimates of Production Function for Different Periods

	1965–70	1970–75	1976–79	1980–86
Observations	58	58	116	203
Constant	4.061	4.286	5.11	6.031
Land	.256	.328	.208	.04
	(7.41)	(6.44)	(6.79)	(2.75)
Labor	.431	.431	.085	.092
	(8.88)	(4.49)	(.798)	(2.72)
Machinery	.050	.110	.488	.554
	(1.64)	(1.48)	(9.71)	(30.27)
C. Fertilizer	.219	.122	.172	.284
	(32.21)	(12.36)	(17.77)	(40.36)
M. Fertilizer	.095	.039	.169	.104
	(7.56)	(1.91)	(9.27)	(13.88)
Region dummy 2	–.348	–.343	–.266	–.316
	(–6.422)	(–4.14)	(–5.25)	(–13.14)
Region dummy 3	–.049	–.118	–.206	–.165
	(–.763)	(–1.26)	(–3.99)	(–8.95)
Region dummy 4	–.108	–.059	–.181	–.078
	(–1.76)	(–.499)	(–2.10)	(–2.37)
Region dummy 5	–.153	–.079	.201	.051
	(–1.78)	(–.522)	(1.92)	(1.49)
Region dummy 6	–.185	–.316	–.069	–.038
	(–1.96)	(–1.93)	(–.536)	(–.788)
Region dummy 7	–.136	–.150	.174	–.006
	(–1.71)	(–.998)	(1.554)	(–.146)
Sum of Input Elasticities	1.051	1.020	1.112	1.038

Appendix E Derivation of Likelihood Function of Frontier Production Function

The derivation of the likelihood follows Aigner et al (1977) closely. Consider the simple model

$$y_i = b'x_i + \varepsilon_i, \qquad i = 1, \ldots, N.$$

where $\varepsilon_i = v_i + u_i$, and v_i represents the symmetric disturbance: i.e., the v_i is assumed to be independently and identically distributed as $N(0,\sigma_v^2)$. The u_i is assumed to be distributed indepently of v_i, and $u_i \leq 0$, and u has a normal distribution above zero.

The distribution of ε which is the sum of a symmetric normal random variable and a truncated normal random variable was derived by M.A. Weinstein in 1964. The density function of ε is:

$$f(\varepsilon) = \frac{2}{\sigma} f^*\left(\frac{\varepsilon}{\sigma}\right)[1-F^*(\varepsilon\lambda\sigma^{-1})], \quad -\infty \leq \varepsilon \geq \infty$$

where $\sigma^2 = \sigma_u^2 + \sigma_v^2$, $\lambda = \sigma_u/\sigma_v$, and $f^*(.)$ and $F^*(.)$ are the standard normal density and distribution functions, respectively. The density of ε is asymmetric around zero, with its mean and variance given by

$$E(\varepsilon) = E(u) = -\frac{\sqrt{2}}{\sqrt{\pi}}\sigma_u$$

$$V(\varepsilon) = V(u) + V(v) = \left(\frac{\pi-2}{\pi}\right)\sigma_u^2 + \sigma_v^2,$$

We assume the sample has N observations, then the log-likelihood function is

$$\ln L(y/b,\lambda,\sigma^2) = N\ln \frac{\sqrt{2}}{\sqrt{\pi}} + N\ln\sigma^{-1} + \sum_{i=1}^{N} \ln[1-F^*(\varepsilon_i \lambda\sigma^{-1})]$$

$$- \frac{1}{2\sigma^2} \sum_{i}^{N} \varepsilon i^2$$

Taking derivatives,

$$\frac{\partial \ln L}{\partial \sigma^2} = - \frac{N}{2\sigma^2} + \frac{1}{2\sigma^4} \sum_{i=1}^{N}(y_i - bx'_i) + \frac{1}{2\sigma^2} \sum_{i=1}^{N} \frac{f_i^*}{(1-F_i^*)} (y_i - bx'_i),$$

$$\frac{\partial \ln L}{\partial \lambda} = - \frac{1}{\sigma} \sum_{i=1}^{N} \frac{f_i^*}{(1-F_i^*)} (y_i - bx'_i),$$

$$\frac{\partial \ln L}{\partial b} = \frac{1}{\sigma^2} \sum_{i=1}^{N} (y - bx'_i) x_i + \frac{\lambda}{\sigma} \sum_{i=1}^{N} \frac{f_i^*}{(1-F_i^*)} x_i$$

where $xk_i \times 1$)vector, and f_i^* and F_i^* are the stabdard norma densitystribution function evaluated at $(y_i^* - b'x_i)\lambda\sigma^{-1}$.

The Second-order derivatives are derived as follow:

$$\frac{\partial^2 \ln L}{\partial \lambda^2} = \frac{1}{\sigma^2} \sum_{i=1}^{N} x_i' x_i + \frac{\lambda^2}{\sigma^2} \sum_{i=1}^{N} \frac{f_i^*}{(1-F_i^*)^2} \Big(-f_i^* + \frac{\lambda}{\sigma}(1-F_i^*)$$

$$(y_i - b'x_i)\Big) x_i' x_i,$$

$$\frac{\partial^2 \ln L}{\partial (\sigma^2)^2} = \frac{1}{2\sigma^4} - \frac{1}{\sigma^6} \sum_{i=1}^{N} (y_i - b'x_i)^2 + \frac{\lambda}{4\sigma^5} \sum_{i=1}^{N} \frac{f_i^*}{(1-F_i^*)^2}$$

$$\Big(- \frac{\lambda}{\sigma} f_i^* (y_i - b'x_i)^2 + \frac{\lambda^2}{\sigma^2} (1-F_i^*)(y_i - b'x_i)^3$$

$$-3(1-F_i^*)(y_i - b'x_i)\Big),$$

$$\frac{\partial^2 \ln L}{\partial \lambda \partial b} = - \frac{1}{\sigma} \sum_{i=1}^{N} \frac{f_i^*}{(1-F_i^*)} (-(1-F_i^*) - \frac{\lambda}{\sigma} f_i^* (y_i - b_i' x)$$

$$+ \frac{\lambda^2}{\sigma^2} (1-F_i^*)(y_i - b'x_i)^2\Big) x_i,$$

$$\frac{\partial^2 \ln L}{\partial \sigma^2 \partial b} = \frac{1}{\sigma^4} \sum_{i=1}^{n} (y_i - b'x_i) x_i + \frac{\lambda}{2\sigma^3} \sum_{i=1}^{N} \frac{f_i^*}{(1-F_i^*)^2} \Big(- (1-F_i^*)$$

$$- \frac{\lambda}{\sigma} f_i^* (y_i - b'x_i) + \frac{\lambda^2}{\sigma^2} (1-F_i^*)(y_i - b'x_i)^2\Big) x_i,$$

$$\frac{\partial^2 \ln L}{\partial \sigma^2 \partial \lambda} = \frac{1}{2\sigma^2} \sum_{i=1}^{N} \frac{f_i^*}{(1-F_i^*)^2} \Big((1-F_i^*)(y_i-b'x_i) + \frac{\lambda}{\sigma} f_i^*(y_i-b'x_i)^2$$

$$-\frac{\lambda}{\sigma^2}(1-F_i^*)(y_i-b'x_i)^3\Big).$$

BIBLIOGRAPHY

Afrait, S. N. 1972 "Efficiency Estimation of Production Functions." International Economic Review, 13(3): 568-598.

Agricultural Yearbook Editing Committee. 1980, 1981, 1982, 1983, 1984, 1985, 1986, and 1987. China Agricultural Yearbook. Beijing: Agricultural Publishing House

Aigner, D. J., and S. F. Chu. 1968. "On the Estimating the Industry Production Function." American Economic Review, 58(4): 826-839.

Aigner, D. J., C. A. K. Lovell, and P. J. Schmidt. 1977. "Formulation and Estimation of Stochastic Frontier Production Function Models." Journal of Econometrics, 6(July): 21-37.

Antle, J. M. 1984. "The Structure of U.S. Agricultural Technology, 1910-78." American Journal of Agricultural Economics, 66(4): 414-421.

Barker, Randolph, Radha Sinha and Beth Rose. 1982. The Chinese Agricultural Economy. Boulder: Westview Press,Inc.

Battese, Geoge E. and Tim J. Coelli. 1988. "Prediction of Firm-Level Technical Efficiencies with a generalized Frontier Production Function and Panel Data." Journal of Econometrics, 38: 387-399.

Bhattacharjee, Jyoti P. 1955. "Resource Use and Productivity In World Agriculture." Journal of Farm Economics, 37: 57-71.

Binswanger, Hans P. and Vernon W. Ruttan. 1978. Induced Innovation -- Technology, Institutions and Development. Baltimore: The Johns Hopkins University Press.

Brada C. Josef and Karl-Eugen Wadokin. 1988. Transition of Socialist Agriculture. Boulder: Westview Press.

Charnes, A., W. W. Cooper, and E. Rhodes. 1978. "Measuring the Efficiency of Decision-Making Units." European Journal of Operational Research, 2(6): 429-444.

Chavas, Jean-Paul and Thomas C. Cox. 1987. "A Non-parametric Analysis of Productivity: The Case of U.S. Agriculture." Staff Paper #281, Department of Agricultural Economics, University of Wisconsin.

Chen, Dunyi, and Jishan Hu. 1983. Chinese Economic Geography. Beijing: China Perspective Publishing House.

China, State Statistical Bureau, Agricultural Statistics Division Editor. 1985, 1986, and 1987. China Rural Statistical Yearbook. Beijing: Statistical Publishing House.

China, State Statistical Bureau, Editorial Office.1985, 1986, 1987, and 1988. Statistical Abstract. Beijing: Statistical Publishing House.

China, State Statistical Bureau, Editorial Office. 1981, 1982, 1983, 1984, 1985, 1986, and 1987. China Statistical Yearbook. Beijing: Statistical Publishing House.

Chinese Academy of Agricultural Science. 1984. Regionalization of Crop Production in China. Beijing: Agricultural Publishing House.

Chinese Academy of Science. 1980. A General Treatise of Chinese Agricultural Geography. Beijing: The Science Publishing House.

Christensen, L. R. 1975. "Concepts and Measurement of Agricultural Productivity." American Journal of Agricultural Economics, 57: 910-915.

Cowing, G. Thomas and Rodney E. Stevenson. 1981. Productivity Measurement in Regulated Industries. New York: Academic Press.

Danilin, V. I. and Ivan S. Materov. 1985. "Measuring Enterprise Efficiency in the Soviet Union: A Stochastic Frontier Analysis." Economica, 52: 225-233.

Denny, M. and M. Fuss. 1983. "A General Approach to Intertemporal and Interspatial Productivity Comparison." Journal of Econometrics 23: 315-330.

Diewert, W. E. 1982. "Duality Approaches to Microeconomic Theory", in K. J. Arrow and M. D. Intriligator (eds), Handbook of Mathematical Economics Volume II. Amsterdam: North-Holland.

Diewert, W. E. 1976. "Exact and Superlative Index Numbers." Journal of Econometrics, 4: 115-146.

Division of Economic Geography, The Institute of Geography, Chinese Academy of Science. 1983. The Distribution of Agricultural Production in China. Beijing: Agricultural Publishing House.

Dogramaci, Ali and Rolf Fåre. 1988. Application of Modern Production Theory: Efficiency and Productivity. Boston: Kluwer Academic Publishers.

Ekanayake S. A. B. and S. K. Jayasuriya. 1987. "Measurement of Firm-specific Technical Efficiency: A Comparison of Methods." Journal of Agricultural Economics Jan. 1987.

Farrell, M. J. 1957. "The Measurement of Productive Efficiency." Journal of the Royal Statistical Society, Ser. A, General, 120, pt. 3. 253-281.

Forsund, F. R., and L. Hjalmarsson. 1979. "Frontier Production Functions and Technical Progress: A Study of General Milk Processing in Swedish Dairy Plants." Econometrica 47(4): 893-900

Forsund, F. R. and E. S. Jansen. 1977. "On Estimating Average and Best Practice Homothetic Production Functions via Cost Functions." International Economic Review 18(2): 463-476.

Fuss, Melvyn and Daniel McFadden. 1978. Production Economics: A Dual Approach to Theory and Application. Vol. 1 and 2. North-Holland Publishing Company.

Greene, W. H. 1980. "Maximum Likelihood Estimation of Econometric Frontier Production Functions." Journal of Econometrics 13 (May): 27:56.

Griliches, Zvi. 1957. "Specification Bias in Estimates of Production Functions." Journal of Farm Economics 39(1): 8-20.

Handbook of Economic For Technology (Agriculture) Editing Committee. 1983. Handbook of Economic for Technology (Agriculture). Beijing: Agricultural Publishing House.

Hayami, Yujiro. 1975. A Century of Agricultural Growth in Japan. Minneapolis: University of Minnesota Press.

Hayami, Yujiro and Vernon W. Ruttan. 1971/1985. Agricultural Development: An International Perspective. Baltimore: The Johns Hopkins University Press.

Hsiao, Cheng. 1986. Analysis of Panel Data. Cambridge: Cambridge University Press.

Hsu, Robert C. 1982. Food for One Billion -- China's Agriculture Since 1949. Boulder: Westview Press.

Huang, C. J. and F. S. Bagi. 1984. "Technical Efficiency on Individual Farms in Northwest India." Southern Economic Journal 51(1):108-115.

Ishikawa, Shigeru. 1981. Essays on Technology Employment and Institutions in Economic Development. The Institute of Economic Research, Hitotsubashi University.

Jorgenson, D. W. and Z. Griliches. 1971. " Divisa Index and Productivity Measurement." Review of Income and Wealth 17: 227-229

Kalirajam. K. P. and Flinn J. C. 1983. "The Measurement of Farm Specific Technical Efficiency." Pakistan Journal of Applied Economics, 2(2). 167-180.

Kendrick, John W. and Beatrice N. Vaccara. 1980. New Developments in Productivity Measurement and Analysis. Chicago: The University of Chicago Press.

Lardy, R. Nicholas. 1983. Agriculture in China's Modern Economic Development. Cambridge: Cambridge University Press.

Lau, J. Lawrence and Pan A. Yotopoulos. 1987. "The Meta-Production Function Approach to Technological Change in World Agriculture." Seminar Paper Presented at the University of Minnesota.

Lee, L. F. 1983. "A Test for Distributional Assumption for the Stochastic Frontier Functions." Journal of Econometrics, 22(August): 245-267

Lee, L. F. and W. G. Tyler. 1978. "The Stochastic Frontier Production Function and Average Efficiency: An Empirical Analysis." Journal of Econometrics 7(June): 385-389.

Leibenstein, H. 1965. "Allocative Efficiency vs. X-efficiency." American Economic Review, 56: 392-415.

Lin, Justin Yifu. 1988. " The Household Responsibility System in China's Agricultural Reform: A Theoretical and Empirical Study." Economic Development and Cultural Change 36(3): 200-224.

Lovell, C.A.K., and Schmidt. 1988. "A Comparison of Alternative Approaches to the Measurement of Productive Efficiency." In Dorgramaci, Ali and Rolf Fare (ed), Application of Modern Production Theory: Efficiency and Productivity. Boston: Kluwer Academic Publishers.

Major Events in Chinese Agricultural Editing Committee. 1981 and 1984. Major Events in Chinese Agriculture. Beijing: Agricultural Publishing House.

Meeusen, W, and J. van den Broeck. 1977. "Efficiency Estimation from Cob-Douglas Production Functions with Composed Error." International Economic Review, 18(2): 435-444.

Mellor John W. 1966. The Economics of Agricultural Development. Ithaca: Cornell University Press.

Murray, Brown. 1966. On the Theory and Measurement of Technological Change. Cambridge: Cambridge University Press.

Nishimizu, Mieko and John M Page, Jr. 1982. "Total Factor Productivity Growth, Technological Progress and Efficiency Change: Dimension of Productivity Change in Yugoslavia, 1965-78." Economic Journal, 92: 920-936.

Pannell, W. Glifton and J. C. Lawrence Ma. 1983. China: The Geography of Development and Modernization. V. H. Winston and Sons.

Perkins, Dwight. 1969. Agricultural Development in China 1368-1968. Chicago: Aldine Publishing Company.

Perkins, Dwight and Shahid Yusuf. 1984. Rural Development in China. Baltimore: The Johns Hopkins University Press.

Peterson, Willis and Yujiro Hayami. 1977. "Technical Change in Agriculture", in Lee Martin (ed.), A Survey of Agricultural Economics Literature. (Vol. 1). Minneapolis: University of Minnesota Press, 497-540.

Pitt, M. M. and L. F. Lee. 1981. "The Measurement and Sources of Technical Inefficiency in the Indonesian Weaving Industry." Journal of Development Economics 9: 43-64.

Rawski, Thomas. 1982. "Agricultural Employment and Technology." in Randolph Barker, Radha Sinha, and Beth Rose (eds.), The Chinese Agricultural Economy. Boulder: Westview Press.

Richmond, J. 1974. "Estimating the Efficiency of Production." International Economic Review 15(2): 515-521.

Ruttan, Vernon W. 1982. Agricultural Research Policy. Minneapolis: University of Minnesota Press.

Ruttan, Vernon W. 1956. "The Contribution of Technological Progress to Farm Output, 1950-1975." Review of Economics and Statistics 38: 61-69.

Ruttan, Vernon W. 1954. "Technological Progress in the Meat Packing Industry." USDA Marketing Res. Rep. 59.

Schultz. T. W. 1964. Transforming Traditional Agriculture. New Haven: Yale University Press.

Solow, Robert. 1957. "Technical Change and the Aggregate Production Function". Review of Economics and Statistics, 39(August): 312-322.

Stevens, Robert D. 1977. Traditional and Dynamics in Small-Farm Agriculture Ames: The Iowa State University Press.

Sun, Han. 1987. "9.8 Million Square Kilometers of Territory." in Wittwer Sylvan (eds), Feeding A Billion. East Lansing: Michigan State University Press.

Tam, On Kit. 1985. China's Agricultural Modernization. Dover, New Hampshire: Groom Helm.

Tang, Anthony and Bruce Stone. 1980. Food Production In PRC. Washington D.C.: International Food Policy Research Institute.

Tang, Anthony. 1984. An Analytical and Empirical Investigation of Agriculture in Mainland China, 1952-1980. Taibei: Chung-Hua Institution for Economic Research.

Timmer, C. 1971. "Using a Probabilistic Frontier Production Function to Measure Technical Efficiency." Journal of Political Economy 79: 176-194.

Whitesell, Robert. 1988. " Estimation of Output Loss from Allocative Inefficiency: Comparison of the Soviet Union and the U.S." Research Menoraudum RM-109, The Center for Development Economics.

Wiens, Thomas B. 1982. "Technical Change", in Randolph Barker et al. (eds.), The Chinese Agricultural Economy.Boulder: Westview Press.

Wilkin, Jerzy. 1988. "The Induced Innovation Model of Agricultural Development and the Socialist Economic System." Euro. R. Agr. Eco.15: 211-219.

Wittwer, Sylvan, Youtai Yu, Han Sun, and Lianzhang Wang. 1987. Feeding A Billion. East Lansing: Michigan State University Press.

Wong, Lung-Fai. 1986. Agricultural Productivity in the Socialist Countries. Boulder: Westview Press Inc.

Wong, Lung-Fai and Vernon W. Ruttan. 1988. "Sources of Differences in Agricultural Productivity Growth Among Socialist Countries", in Dogramaci, Ali and Rolf Fare, Application of Modern Production Theory: Efficiency and Productivity. Boston: Kluwer Academic Publishers.

World Bank. 1982. China: Socialist Economic Development. Volume 1, The Economy, Statistical System, and Basic Data. Washington.

World Bank. 1982. China: Socialist Economic Development. Annex c: Agricultural Development. Washington, D. C.

Yamada, Saburo. 1975. A Comparative Analysis of Asian Agricultural Productivities and Growth Patterns. Tokyo:Asian Productivity Organization, Productivity Series No.10.

Yamada, Saburo and Vernon W. Ruttan. 1980. "International Comparison of Productivity in Agriculture." New Development in Productivity Measurement and Analysis. ed. et. John W. Kendrick and Beatrice N. Vaccara. Chicago: The University of Chicago Press, 509-594.

INDEX